T0345141

Linear Algebra With Machine Learning and Data

This book takes a deep dive into several key linear algebra subjects as they apply to data analytics and data mining. The book offers a case study approach where each case will be grounded in a real-world application.

This text is meant to be used for a second course in applications of Linear Algebra to Data Analytics, with a supplemental chapter on Decision Trees and their applications in regression analysis. The text can be considered in two different but overlapping general data analytics categories: clustering and interpolation.

Knowledge of mathematical techniques related to data analytics and exposure to interpretation of results within a data analytics context are particularly valuable for students studying undergraduate mathematics. Each chapter of this text takes the reader through several relevant case studies using real-world data.

All data sets, as well as Python and R syntax, are provided to the reader through links to Github documentation. Following each chapter is a short exercise set in which students are encouraged to use technology to apply their expanding knowledge of linear algebra as it is applied to data analytics.

A basic knowledge of the concepts in a first Linear Algebra course is assumed; however, an overview of key concepts is presented in the Introduction and as needed throughout the text.

Dr. Crista Arangala is Professor of Mathematics and Chair of the Department of Mathematics and Statistics at Elon University in North Carolina. She has been teaching and researching in a variety of fields including inverse problems, applied partial differential equations, applied linear algebra, mathematical modeling and service learning education. She runs a traveling science museum with her Elon University students in Kerala, India. Dr. Arangala was chosen to be a Fulbright Scholar in 2014 as a visiting lecturer at the University of Colombo where she continued her projects in inquiry learning in Linear Algebra and began working with a modeling team focusing on Dengue fever research. Dr. Arangala has published several textbooks that implores inquiry learning techniques including *Exploring Linear Algebra: Labs and Projects with Matlab* and *Mathematical Modeling: Branching Beyond Calculus*.

Textbooks in Mathematics
Series editors:
Al Boggess, Kenneth H. Rosen

https://www.routledge.com/Textbooks-in-Mathematics/book-series/CANDHTEXBOOMTH

Linear Algebra
With Machine
Learning and Data

Crista Arangala

CRC Press
Taylor & Francis Group
Boca Raton London New York

CRC Press is an imprint of the
Taylor & Francis Group, an **informa** business

A CHAPMAN & HALL BOOK

First edition published 2023
by CRC Press
6000 Broken Sound Parkway NW, Suite 300, Boca Raton, FL 33487-2742

and by CRC Press
4 Park Square, Milton Park, Abingdon, Oxon, OX14 4RN

CRC Press is an imprint of Taylor & Francis Group, LLC

© 2023 Taylor & Francis Group, LLC

Library of Congress Cataloging-in-Publication Data

Names: Arangala, Crista, author.
Title: Linear algebra with machine learning and data / Crista Arangala.
Description: First edition. | Boca Raton : CRC Press, 2023. | Series:
Textbooks in mathematics | Includes bibliographical references and
index.
Identifiers: LCCN 2022047264 | ISBN 9780367458393 (hardback) | ISBN
9781032458649 (paperback) | ISBN 9781003025672 (ebook)
Subjects: LCSH: Algebras, Linear--Textbook. | Machine
learning--Mathematics--Textbooks. | Data mining--Mathematics--Textbooks.
Classification: LCC QA184.2 .A73 2023 | DDC 518/.43--dc23/eng20230203
LC record available at https://lccn.loc.gov/2022047264

ISBN: 978-0-367-45839-3 (hbk)
ISBN: 978-1-032-45864-9 (pbk)
ISBN: 978-1-003-02567-2 (ebk)

DOI: 10.1201/9781003025672

Typeset in CMR10
by KnowledgeWorks Global Ltd.

Publisher's note: This book has been prepared from camera-ready copy provided by the authors.

Contents

Acknowledgments

I want to thank the faculty of Elon University and in particular the Elon Mathematics and Statistics Department for always being on the edge of innovative teaching and embracing important initiatives around exposing students in data competency. Many students have also contributed to the success of this project through undergraduate research and classroom experiences. I wish to thank Amy Moore, Abby Weber, Matthew Davenport, Rachel Dietert, Katie Everitt, Lauren Hanchar, Mary Hermes, Xuan Huynh, Adam Jockle, Gautam Kotwal, Clayton McLamb, Maia Sirabian, Sarina Smith, William Tunis, Ashley Vann, and Parker Verlander.

In addition, I would like to recognize the support for this text from Elon University's Faculty Research and Development program.

Preface

This text is meant to be used for a second course in applications of Linear Algebra to Data Analytics, with a supplemental chapter on Decision Trees and their applications in regression analysis. Throughout the text a basic knowledge of the concepts in a first Linear Algebra course is assumed; however, an overview of key concepts is presented in the Introduction and as needed throughout the text.

Knowledge of mathematical techniques related to data analytics, and exposure to interpretation of results within a data analytics context, are particularly valuable for students studying undergraduate mathematics. Each chapter of this text takes the reader through several relevant case studies using real world data. Some case study applications address gerrymandering issues, modeling the spread of disease, digital facial recognition, and racial inequities in higher education. Readers are encouraged to replicate results from the presented case studies. All data sets, as well as Python and R syntax, are provided to the reader through links to Github documentation. Following each chapter is a short exercise set in which students are encouraged to use technology to apply their expanding knowledge of linear algebra as it is applied to data analytics.

The text can be considered in two different but overlapping general data analytics categories: clustering and interpolation. Chapters 1 and 3 focus on eigenvalues and singular values, as well as their associated vectors, that are of key importance in clustering techniques, such as Principal Component Analysis. Chapters 4 through 6 focus on techniques specific to interpolation models. Integrated throughout this text, the reader will also find many machine learning techniques of interest, such as hidden Markov chains in Chapter 2, neural networks in Chapter 5, decision trees and forests in Chapter 6, and random matrices in Chapter 7.

Introduction

In this introduction, we give an overview of background expectations. Also, for those who are looking to use this book for a first course in Mathematical Methods in Data Analytics, an overview of necessary Linear Algebra concepts is presented here and embedded throughout the text.

The entire text is focused on applications of Linear Algebra in Data Analytics, so we begin with a quick overview of matrix algebra. Matrix A has entry $a_{i,j}$ in row i and column j.

$$
A = \begin{pmatrix}
a_{1,1} & a_{1,2} & a_{1,3} & \cdots & a_{1,m} \\
a_{2,1} & a_{2,2} & a_{2,3} & \cdots & a_{2,m} \\
\vdots & \ddots & \ddots & \ddots & \vdots \\
a_{n,1} & a_{n,2} & a_{n,3} & \cdots & a_{n,n}
\end{pmatrix}.
$$

The dimensions of A are $n \times m$. A matrix is **square** if it has the same number of rows as it does columns.

Matrix addition is entry-wise. That is, that if A and B are matrices with the same dimensions, then $(A + B)_{i,j} = a_{i,j} + b_{i,j}$.

If one wishes to multiply each entry of a matrix by a constant, also called a scalar, this is referred to as scalar multiplication.

The transpose of A, denoted A^T, is an $m \times n$ matrix where

$$
A^T_{i,j} = A_{j,i}.
$$

A matrix A is **symmetric** if $A = A^T$.

The **Hadamard product**, also known as the entry-wise product, of two matrices of the same dimensions, A and B, is

$$
A \circ B,
$$

where the $(i,j)^{\text{th}}$ entry of $A \circ B$ is $a_{i,j} b_{i,j}$.

In order to multiply a $m \times n$ matrix A on the right by a matrix B, the number of rows of B must be equal to the number of columns of matrix A. That is B must be a $n \times p$ matrix for some natural number p.

$$
A_{\boxed{m} \times n} \cdot B_{n \times \textcircled{p}} = C_{\boxed{m} \times \textcircled{p}}
$$

The entries of the product of AB

$$AB_{i,j} = \sum_{k=1}^{n} a_{i,k} b_{k,j}.$$

Example 0.1. *It is important to note that matrix multiplication is not commutative. For example, if*

$$A = \begin{pmatrix} 1 & 0 & -2 \\ 4 & 1 & 1 \end{pmatrix} \text{ and } B = \begin{pmatrix} 2 & -1 \\ 0 & 1 \\ 1 & 3 \end{pmatrix},$$

Then,

$$AB = \begin{pmatrix} 0 & -7 \\ 9 & -1 \end{pmatrix}$$

and

$$BA = \begin{pmatrix} -2 & -1 & -5 \\ 4 & 1 & 1 \\ 13 & 3 & 1 \end{pmatrix}.$$

The **identity matrix**, I, is the a square matrix such that

$$I_{i,j} = 1 \text{ if } i = j, \text{ and } 0 \text{ otherwise.}$$

If A is an $m \times n$ matrix and I is the $m \times m$ identity matrix then

$$AI = IA = A.$$

If A is a square matrix, we can also talk about the powers of A, where $A^2 = A \cdot A$ and exponential rules of matrices hold.

$$A^{r+s} = A^r \cdot A^s$$
$$(A^r)^s = A^{rs}$$

In many applications of Linear Algebra, the goal of the problem is to solve for x in the system $Ax = b$, where A is an $n \times n$ matrix and x and b are $n \times 1$ matrices, also called vectors in $\mathbb{R}^{\kappa}$.

If A is **invertible**, also called **non-singular**, then there exists a matrix A^{-1} such that

$$A^{-1} \cdot A = A \cdot A^{-1} = I$$

If A is not invertible, then we say that A is **singular**.

A system of m equations in n variables

$$a_{1,1}x_1 + a_{1,2}x_2 + \cdots + a_{1,n}x_n = b_1$$
$$a_{2,1}x_1 + a_{2,2}x_2 + \cdots + a_{2,n}x_n = b_2$$
$$\vdots$$
$$a_{m,1}x_1 + a_{m,2}x_2 + \cdots + a_{m,n}x_n = b_m$$

can be written in the matrix form $Ax = b$ where A is a $m \times n$ coefficient matrix and x and b are $m \times 1$ matrices.

If A is square and invertible then the system $Ax = b$ has exactly one solution, $x = A^{-1}b$. If A is not square or is not invertible, then it is possible for the system have either no solutions or infinitely many solutions. It is also important to note that if $Ax = \vec{0}$ and A is invertible then the only solution is $x = \vec{0}$, also called the **trivial solution**.

The **determinant** of a square matrix A maps a matrix to a real number, $|A|$. There are a variety of techniques for calculating the determinant of a matrix, including cofactor expansion and formulas for small matrices such as 2×2 matrices.

$$\begin{vmatrix} a_{1,1} & a_{1,2} \\ a_{2,1} & a_{2,2} \end{vmatrix} = a_{1,1}a_{2,2} - a_{1,2}a_{2,1}.$$

In order to perform **cofactor expansion** on a $n \times n$ matrix, A, where $n > 2$, choose a row, i, or column, j to expand on. Then

$$|A| = \sum_{j=1}^{n}(-1)^{i+j}a_{i,j}M_{i,j} = \sum_{i=1}^{n}(-1)^{i+j}a_{i,j}M_{i,j},$$

where $M_{i,j}$ is the $(i,j)^{\text{th}}$ **minor** and is the determinant of the $(n-1) \times (n-1)$ submatrix of A created by deleting the ith row and j^{th} column of A.

Example 0.2.

$$\begin{vmatrix} 1 & 2 \\ 3 & 4 \end{vmatrix} = 1 \cdot 4 - 2 \cdot 3 = -2$$

$$\begin{vmatrix} 2 & 0 & 1 \\ 0 & 1 & 2 \\ 0 & 3 & 4 \end{vmatrix} = (-1)^2 \cdot 2 \begin{vmatrix} 1 & 2 \\ 3 & 4 \end{vmatrix} + (-1)^3 \cdot 0 \begin{vmatrix} 0 & 2 \\ 0 & 4 \end{vmatrix} + (-1)^4 \cdot 1 \begin{vmatrix} 0 & 1 \\ 0 & 3 \end{vmatrix} = -4$$

when we expand on the first row. Note that one can expand on any row or column, but if you are attempting to be efficient choose a row or column with the most zeros.

Most importantly, the fact that determining if a square matrix is invertible is equivalent to determining if $|A| \neq 0$.

Chapter 1 focuses on the applications of Linear Algebra to Graph Theory, where powers of matrices, bases, eigenvalues, and eigenvectors are heavily implemented. Eigenvalues and eigenvectors will be reviewed in Chapter 1. A brief review of bases is presented here.

A set of vectors $S = \{v_1, v_2, \ldots, v_n\}$ is called **linearly independent** if the equation

$$k_1 v_1 + k_2 v_2 + \cdots + k_n v_n = 0$$

has only the trivial solution, $k_1 = k_2 = \cdots = k_n = 0$. If the set is not linearly independent, we called the set **linearly dependent**.

If $v_1, v_2, \ldots, v_n$ are vectors in $\mathbb{R}^n$ and are the columns of a matrix A, then

$$A = \begin{pmatrix} | & | & \cdots & | \\ | & | & \cdots & | \\ v_1 & v_2 & \cdots & v_n \\ | & | & \cdots & | \\ | & | & \cdots & | \end{pmatrix}$$

$$A \begin{pmatrix} k_1 \\ k_2 \\ \vdots \\ k_n \end{pmatrix} = \begin{pmatrix} 0 \\ 0 \\ \vdots \\ 0 \end{pmatrix}.$$

has only the trivial solution if A is invertible.

If $S = \{v_1, v_2, \ldots, v_n\}, v_i \in \mathbb{R}^n$, then we say that $w \in \mathbb{R}^n$ can be written as a **linear combination** of the vectors in S if there exists $k_1, k_2, \ldots, k_n$ such that

$$w = k_1 v_1 + k_2 v_2 + \cdots + k_n v_n.$$

We say that S **spans** $\mathbb{R}^n$ if every vector $w \in \mathbb{R}^n$ can be written as a linear combinations of the vectors in S.

Example 0.3. $S_1 = \{(1,0,0), (0,1,0), (0,0,1)\}$, $S_2 = \{(1,2,0), (0,3,4), (0,0,5)\}$, and $S_3 = \{(1,1,0), (0,0,1), (0,0,0)\}$ *are sets of vectors in* $\mathbb{R}^3$ *and*

$$A_1 = \begin{pmatrix} 1 & 0 & 0 \\ 0 & 1 & 0 \\ 0 & 0 & 1 \end{pmatrix}, A_2 = \begin{pmatrix} 1 & 0 & 0 \\ 2 & 3 & 0 \\ 0 & 4 & 5 \end{pmatrix}, A_3 = \begin{pmatrix} 1 & 1 & 0 \\ 1 & 0 & 0 \\ 0 & 1 & 0 \end{pmatrix}.$$

Note that $|A_1| = 1$, $|A_2| = 15$, and $|A_3| = 0$ and thus matrices A_1 and A_2 are non-singular and A_3 is singular. Thus S_1 and S_2 are linearly independent sets that span $\mathbb{R}^3$.

The set S_3 is linearly dependent and does not span $\mathbb{R}^3$. You can see that S_3 is linearly dependent since the equation

$$k_1(1,1,0) + k_2(0,0,1) + k_3(0,0,0) = (0,0,0),$$

has solutions other than the trivial solution, for example $k_1 = k_2 = 0$, $k_3 = 1$. In order for S_3 to span $\mathbb{R}^3$, every vector (x,y,z) in $\mathbb{R}^3$ must be able to be written as a linear combination of the vectors in S_3.

If you remove a vector from S_1, for example $\{(1,0,0), (0,1,0)\}$, the set is still linearly independent but does not span $\mathbb{R}^3$.

If you add a vector to S_1, for example $\{(1,0,0), (0,1,0), (0,0,1), (1,2,3)\}$, the set is linearly dependent but still spans $\mathbb{R}^3$.

If a set S is a set of n linearly independent vectors which spans $\mathbb{R}^n$ then S is a **basis** for $\mathbb{R}^n$.

Example 0.4. *The standard basis for $\mathbb{R}^3$ is the row vectors of the 3×3 identity matrix*

$$\{(1,0,0), (0,1,0), (0,0,1)\};$$

however, this is not the only basis for $\mathbb{R}^3$. A set of any three vectors in $\mathbb{R}^3$ that are linearly independent form a basis for $\mathbb{R}^3$.

Similarly, the span of the set $S_1 = \{(2,1)\}$ are the vectors in $\mathbb{R}^2$ where the value of the x-coordinate is twice the value of the y-coordinate. Therefore, $S_1 = \{(2,1)\}$ is a basis for the set

$$V = \{(2x,x)| \ x \in \mathbb{R}\}.$$

Other sets such as $S_2 = \{(4,2)\}$ and $\{(2\pi,\pi)\}$ are also bases for V.

The main focus of Chapter 2, Stochastic Processes, is Markov Chains. Some basic probability theory and knowledge of eigenvalues and eigenvectors would be particularly useful in this chapter. Related to probability theory, it is helpful for the reader to have some understanding of probability theory, particularly the relationship between joint and conditional probability. For a basic overview of probability theory, one can refer to Section 5.1 prior to reading Chapter 2. For the most part, one could read Chapter 2 independently of Chapter 1; however, the beginning stages of Chapter 2 do rely on basic knowledge of graph theory.

Chapter 3 focuses on the implementation of Singular Value Decomposition and Principal Component Analysis in the context of data science. A good understanding of eigenvalues and eigenvectors is instrumental in this chapter. An overview of inner product spaces and orthogonality is provided. This particular chapter also becomes fairly technology heavy with several exercises relying on

R or Python knowledge. Again, Chapter 3 can be taught, for the most part, independently of the previous 2 chapters, although further chapters benefit from the knowledge of Chapter 3 material.

Both Chapters 4 and 5 focus on predictive models. Where Chapter 4 focuses more on traditional interpolation techniques such as Chebyshev, Hermite, and Lagrange interpolation, Chapter 5 steps back and takes a quick look at probability theory and matrix calculus before connecting the material from Chapter 4 to machine learning and neural network techniques. In these sections, some knowledge of calculus is helpful, since there is a focus on optimizing cost functions. Again, a good understanding of technology is important in Chapter 5, as iterative methods become significantly important in neural network techniques.

Chapter 6 focuses on implementing the concepts of previous chapters in decision trees. Several different takes on decision trees will be presented in this chapter, including decision tree regression, Fourier transformations, and random forests.

Finally, many models are built around random matrices; the focus is on Chapter 7. The discussion in Chapter 7 relies heavily on probability theory, introduced in Chapter 5, and knowledge of eigenvalues and eigenvectors. In Chapter 7, the **Kroenecker product**, also called the **tensor product** is also used.

If A is an $m \times n$ matrix and B is a $p \times q$ matrix then the Kroenecker product,

$$A \otimes B = \begin{pmatrix} a_{1,1}B & \cdots & a_{1,n}B \\ \vdots & & \vdots \\ a_{m,1}B & \cdots & a_{m,n}B \end{pmatrix}.$$

Example 0.5. *If A is a 2×2 matrix,* $\begin{pmatrix} a_{1,1} & a_{1,2} \\ a_{2,1} & a_{2,2} \end{pmatrix}$ *and B is a 3×2 matrix,*

$\begin{pmatrix} b_{1,1} & b_{1,2} \\ b_{2,1} & b_{2,2} \\ b_{3,1} & b_{3,2} \end{pmatrix}$*, then*

$$A \otimes B = \begin{pmatrix} a_{1,1}B & a_{1,2}B \\ a_{2,1}B & a_{2,2}B \end{pmatrix}$$

$$= \begin{pmatrix} a_{1,1}b_{1,1} & a_{1,1}b_{1,2} & a_{1,2}b_{1,1} & a_{1,2}b_{1,2} \\ a_{1,1}b_{2,1} & a_{1,1}b_{2,2} & a_{1,2}b_{2,1} & a_{1,2}b_{2,2} \\ a_{1,1}b_{3,1} & a_{1,1}b_{3,2} & a_{1,2}b_{3,1} & a_{1,2}b_{3,2} \\ a_{2,1}b_{1,1} & a_{2,1}b_{1,2} & a_{2,2}b_{1,1} & a_{2,2}b_{1,2} \\ a_{2,1}b_{2,1} & a_{2,1}b_{2,2} & a_{2,2}b_{2,1} & a_{2,2}b_{2,2} \\ a_{2,1}b_{3,1} & a_{2,1}b_{3,2} & a_{2,2}b_{3,1} & a_{2,2}b_{3,2} \end{pmatrix},$$

Notice that the Kronecker product is not commutative,

$$B \otimes A = \begin{pmatrix} b_{1,1}A & b_{1,2}A \\ b_{2,1}A & b_{2,2}A \\ b_{1,3}A & b_{3,2}A \end{pmatrix}$$

$$= \begin{pmatrix} b_{1,1}a_{1,1} & b_{1,1}a_{1,2} & b_{1,2}a_{1,1} & b_{1,2}a_{1,2} \\ b_{1,1}a_{2,1} & b_{1,1}a_{2,2} & b_{1,2}a_{2,1} & b_{1,2}a_{2,2} \\ b_{2,1}a_{1,1} & b_{2,1}a_{1,2} & b_{2,2}a_{1,1} & b_{2,2}a_{1,2} \\ b_{2,1}a_{2,1} & b_{2,2}a_{2,2} & b_{2,2}a_{2,1} & b_{2,2}a_{2,2} \\ b_{3,1}a_{1,1} & b_{3,1}a_{1,2} & b_{3,2}a_{1,1} & b_{3,2}a_{1,2} \\ b_{3,1}a_{2,1} & b_{3,2}a_{2,2} & b_{3,2}a_{2,1} & b_{3,2}a_{2,2} \end{pmatrix},$$

A matrix A is **skew-symmetric** if $A^T = -A$. In Chapter 7, we specifically employ the Kronecker product of a skew-symmetric matrix with the identity matrix as is seen in Example 0.6.

Example 0.6.

$$\begin{pmatrix} 0 & -1 \\ 1 & 0 \end{pmatrix} \otimes \begin{pmatrix} 1 & 0 \\ 0 & 1 \end{pmatrix} = \begin{pmatrix} 0 & 0 & -1 & 0 \\ 0 & 0 & 0 & -1 \\ 1 & 0 & 0 & 0 \\ 0 & 1 & 0 & 0 \end{pmatrix}.$$

Throughout the text, readers are encouraged to use a critical eye when it comes to ethics of data analytics. Additionally, when choosing a model, the reader should take into account the application in order to chose the most meaningful model in context.

1

Graph Theory

1.1 Basic Terminology

When studying data, data scientists are most concerned with the relationships or connections between the pieces of data in a network. In this chapter, we explore those networks in terms of graph theory and see how these ideas tie back to data science and linear algebra.

A **graph**, G, can be represented by a set of **vertices** or **nodes**, V, and a set of **edges**, E, where each edge in E connects two vertices in V.

The number of vertices in a graph, G, is called the **order** of the graph and is denoted by $|V|$. The number of edges in a graph, G, is called the **size** and is denoted by $|E|$.

An edge is said to be **incident** to vertices v_i and v_j if it connects the two vertices. We call vertices v_i and v_j **endpoints** of this edge.

Vertex v_i is said to be **adjacent** to vertex v_j if there is an edge incident to both vertex v_i and vertex v_j. The **degree** of a vertex is the number of incident edges to that vertex.

Notice that in the graph in Figure 1.1, $|V| = 6$ and $|E| = 12$. Vertex v_4 is adjacent to every other vertex while v_1 is adjacent to v_2, v_3, and v_4, and thus $degree(v_1) = 3$ and $degree(v_4) = 5$.

Theorem 1. *(Handshaking Theorem) In any graph with n vertices, $v_1, v_2, \ldots, v_n$ and m edges,*

$$\sum_{i=1}^{n} degree(v_i) = 2m.$$

An edge with only one endpoint is called a **loop**.

Notice that if a vertex is adjacent to itself via a loop, that edge contributes two edges to the degree of that vertex. The handshaking theorem emphasizes that each edge is incident to two vertices, which may or may not be unique, and is counted twice when calculating the sum of the degrees of the graph.

If two or more edges have the same endpoints, then they are called **multiple edges**.

Figure 1.2 shows a graph with multiple edges between vertices v_1 and v_2 and a loop from vertex v_3 to itself. Vertex v_3 has a degree of 4, where the loop

DOI: 10.1201/9781003025672-1

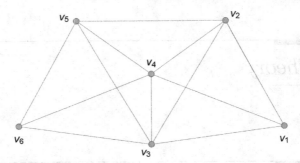

FIGURE 1.1
$|V| = 6, |E| = 12$, degree $(v_1) = 3$, and degree $(v_4) = 5$.

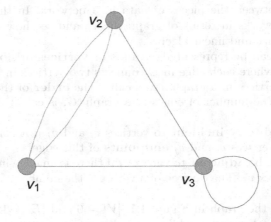

FIGURE 1.2
Example of a graph with a loop and multiple edges.

from v_3 to itself contributes two edges to the degree of v_3. Vertices v_1 and v_2 both have a degree of 3. The multiple edges between v_1 and v_2 contribute 2 edges to the degree of both v_1 and v_2.

A **subgraph** of graph G is another graph containing a subset of vertices from V and all of the edges from E that connect pairs of vertices in the subset.

There are many subgraphs of the graph displayed in Figure 1.1; one example can be seen in Figure 1.3.

For each graph, there is a corresponding **adjacency matrix**, A, where

$$A_{i,j} = \begin{cases} 1, & \text{if vertex } v_i \text{ is adjacent to vertex } v_j, \\ 0, & \text{otherwise.} \end{cases}$$

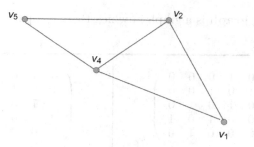

FIGURE 1.3
Example of a subgraph of the graph in Figure 1.1.

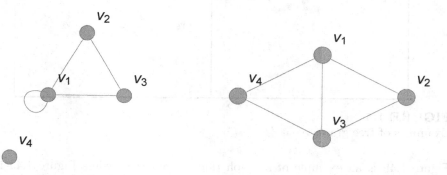

FIGURE 1.4
a. Graph with a loop, b. Graph with no loops.

Notice the loop in Figure 1.4a from vertex v_1 to itself. This is represented in the adjacency matrix A, $a_{1,1} = 1$.

Note also that the vertex v_4 is not adjacent to any vertex in Figure 1.4a and thus every entry of A in row 4 and column 4 has a value of 0.

Figure 1.4b is an example of a graph with no loops in which all vertices are adjacent to at least one other vertex in the graph.

The corresponding adjacency matrices are

$$A = \begin{pmatrix} 1 & 1 & 1 & 0 \\ 1 & 0 & 1 & 0 \\ 1 & 1 & 0 & 0 \\ 0 & 0 & 0 & 0 \end{pmatrix} \quad \text{and } B = \begin{pmatrix} 0 & 1 & 1 & 1 \\ 1 & 0 & 1 & 0 \\ 1 & 1 & 0 & 1 \\ 1 & 0 & 1 & 0 \end{pmatrix}.$$

A **path** from a vertex v_i to a vertex v_j is a sequence of incident edges that connect v_i to v_j. For example, in Figure 1.4b, $v_1v_2v_3$ is a two-step path between v_1 and v_3. Note that a path is also a subgraph of the entire graph.

A **connected graph** is a graph in which there is a path between each pair of vertices.

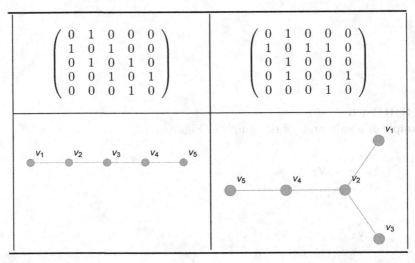

FIGURE 1.5
Examples of tree graphs.

Figure 1.4b is an example of a graph that is connected while Figure 1.4a is disconnected. As we noted earlier, the adjacency matrices affiliated with a disconnected graph contain at least one row and column of zeros.

A **tree graph** is a connected graph in which there is no path between a vertex and itself. Some examples of tree graphs and their corresponding adjacency matrices can be seen in Figure 1.5.

In this text, we will be focusing on **simple connected graphs**, those connected graphs without multiple edges and loops, unless otherwise noted.

All these adjacency matrices, A, presented thus far is **symmetric**, meaning that

$$A = A^T.$$

This is because the edges between each pair of vertices do not have a defined direction.

More formally, a simple connected graph is called **undirected** if, for each pair of adjacent vertices, v_i and v_j, if v_i is adjacent to v_j, then v_j is also adjacent to v_i.

Each of the examples presented in Figures 1.4 and 1.5 are undirected graphs. Figure 1.6 shows an example of a directed graph.

When presented with a directed graph, we talk about the **in-degree** and the **out-degree** of a vertex, where the in-degree is the number of edges coming into the vertex and the out-degree is the number of edges coming out of a vertex.

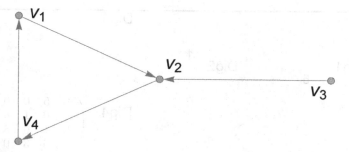

FIGURE 1.6
Example of a directed graph.

In Figure 1.6,

$$\text{in-degree}(v_2) = 2 \text{ and out-degree}(v_2) = 1.$$

Recall that vertex v_i is said to be adjacent to vertex v_j if there is an edge from v_i to v_j. So for example, in Figure 1.6, vertex v_1 is adjacent to vertex v_2; however, vertex v_2 is not adjacent to vertex v_1.

The adjacency matrix associated with the graph in Figure 1.6 is

$$A = \begin{pmatrix} 0 & 1 & 0 & 0 \\ 0 & 0 & 0 & 1 \\ 0 & 1 & 0 & 0 \\ 1 & 0 & 0 & 0 \end{pmatrix}.$$

Notice that A is not a symmetric matrix due to the directed nature of the affiliated graph.

All of the examples thus far have been unweighted graphs; however, a graph's edges can have weights associated with them.

If w_{ij} is the **weight of the edge** between vertex i and vertex j, then the $(i,j)^{\text{th}}$ entry in the adjacency matrix is w_{ij}.

One venue in which graph theory has applications is in archeology and similar fields with artifacts that may share similar traits.

Imagine that there are four digs located near one another and that artifacts are found in each dig. The graph shown in Figure 1.7 shows the digs as vertices and the number of attributes that they share as weights on the edges; this is an example of a weighted graph.

Social networks have also been a hub of activity when it comes to graph theory.

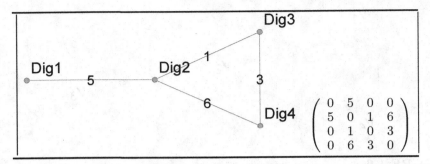

FIGURE 1.7
Example of a weighted graph.

Example 1.1. *Law enforcement has looked to such networks to help fight crime and investigate crime networks. Suppose that the local police have identified Suspects X and Y in Figure 1.8 as individuals involved in a spree of gang-related incidents. However, law enforcement suspects that there are more gang members involved and that they have not identified the leader of the gang network.*

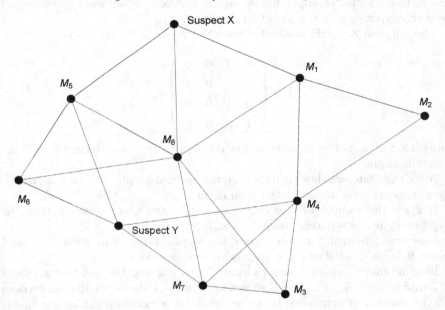

FIGURE 1.8
Adjacency matrix for Example 1.1.

$$A = \begin{array}{c} \\ M_1 \\ M_2 \\ M_3 \\ M_4 \\ M_5 \\ M_6 \\ M_7 \\ M_8 \\ X \\ Y \end{array} \begin{pmatrix} M_1 & M_2 & M_3 & M_4 & M_5 & M_6 & M_7 & M_8 & X & Y \\ 0 & 1 & 0 & 1 & 0 & 0 & 0 & 1 & 1 & 0 \\ 1 & 0 & 0 & 1 & 0 & 0 & 0 & 0 & 0 & 0 \\ 0 & 0 & 0 & 1 & 0 & 0 & 1 & 1 & 0 & 0 \\ 1 & 1 & 1 & 0 & 0 & 0 & 1 & 0 & 0 & 1 \\ 0 & 0 & 0 & 0 & 0 & 1 & 0 & 1 & 1 & 1 \\ 0 & 0 & 0 & 0 & 1 & 0 & 0 & 1 & 0 & 1 \\ 0 & 0 & 1 & 1 & 0 & 0 & 0 & 1 & 0 & 1 \\ 1 & 0 & 1 & 0 & 1 & 1 & 1 & 0 & 1 & 0 \\ 1 & 0 & 0 & 0 & 1 & 0 & 0 & 1 & 0 & 0 \\ 0 & 0 & 0 & 1 & 1 & 1 & 1 & 0 & 0 & 0 \end{pmatrix}$$

For simplicity, in the adjacency matrix, Suspects X and Y will be placed in rows 9 and 10 (and columns 9 and 10), respectively. The adjacency matrix corresponding to the relationships in Figure 1.8, A, shows that both Suspects X and Y are friends with M_5.

In a much larger database, it would be difficult to quickly scan the data to determine how many immediate relationships two vertices, or nodes, in a network have in common. So we will introduce some mathematical ways to explore these relationships here.

The **Gramian matrix,**

$$S = AA^T,$$

can be used to quantify network relationships.

$$S = AA^T = \begin{pmatrix} 4 & 1 & 2 & 1 & 2 & 1 & 2 & 1 & 1 & 1 \\ 1 & 2 & 1 & 1 & 0 & 0 & 1 & 1 & 1 & 1 \\ 2 & 1 & 3 & 1 & 1 & 1 & 2 & 1 & 1 & 2 \\ 1 & 1 & 1 & 5 & 1 & 1 & 2 & 3 & 1 & 1 \\ 2 & 0 & 1 & 1 & 4 & 2 & 2 & 2 & 1 & 1 \\ 1 & 0 & 1 & 1 & 2 & 3 & 2 & 1 & 2 & 1 \\ 2 & 1 & 2 & 2 & 2 & 2 & 4 & 1 & 1 & 1 \\ 1 & 1 & 1 & 3 & 2 & 1 & 1 & 6 & 2 & 3 \\ 1 & 1 & 1 & 1 & 1 & 2 & 1 & 2 & 3 & 1 \\ 1 & 1 & 2 & 1 & 1 & 1 & 1 & 3 & 1 & 4 \end{pmatrix}$$

the main diagonal entries represent the degree of the vertex or in this context the number of connections that the person has in the network.

Suspect X, in row 9, has 3 connections in the network, represented by $S_{9,9}$, and Suspect Y represented in row 10 has 4 connections in the network, as seen in $S_{10,10}$.

Notice also that the $(i,j)^{th}$ entry in S is the number of relationships the ith and j^{th} person have in common.

For example, the entry, $S_{1,3} = 2$, represents the similar network connections between M_1 and M_3. Both of these members are connected to M_4 and M_8.

It is the job of data scientists to look at data and make hypotheses about the future. Another area in which this is particularly applicable is *sports ranking*.

For example, in the first half of the SEC-East 2019 football season, Florida was winning the conference with a 4-1 record. **What if you wished to look at the teams' conference record part-way through the season to predict future rankings?**

Representing the existence of a head-to-head competition in terms of an adjacency matrix may not be the most useful in this case, as the entries of this matrix would be all ones. However, a matrix that could incorporate a win-lose structure as well as the score differential may be beneficial in studying the outcomes of this season.

For example, on September 9, 2019, Florida defeated Kentucky 29 to 21. One might choose to assign a weight of 8, the score differential, to the edge connecting Florida and Kentucky; however, you may want to also be able to indicate the direction of the win.

One way to do this is by using a **directed graph**, where an edge from v_i to v_j may exist without the existence of an edge from v_j to v_i. In this application, a directed edge from vertex v_i to vertex v_j would denote a victory for team i.

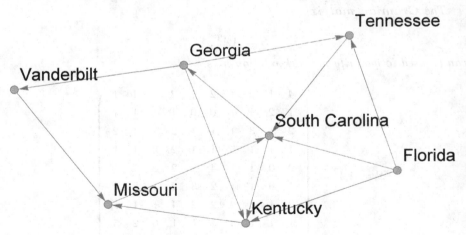

FIGURE 1.9
Directed Graph for SEC-East Football Example.

In the case of a directed graph, the adjacency matrix may not be symmetric. For example, since Florida defeated Kentucky, Florida is adjacent to Kentucky but Kentucky is not adjacent to Florida.

$$
A_{SEC} = \begin{pmatrix}
 & FL & TN & KY & GA & MIZ & SC & VAN \\
FL & 0 & 1 & 1 & 0 & 0 & 1 & 0 \\
TN & 0 & 0 & 0 & 0 & 0 & 1 & 0 \\
KY & 0 & 0 & 0 & 0 & 1 & 0 & 0 \\
GA & 0 & 1 & 1 & 0 & 0 & 0 & 1 \\
MIZ & 0 & 0 & 0 & 0 & 0 & 1 & 0 \\
SC & 0 & 0 & 1 & 1 & 0 & 0 & 0 \\
VAN & 0 & 0 & 0 & 0 & 1 & 0 & 0
\end{pmatrix}
$$

Directed graphs can also be used to visualize a **decision tree** or **neural network** based on raw input data and decision questions that lead to an output. More extensive theory and applications of these techniques can be found in Chapters 6 and 7.

1.2   The Power of the Adjacency Matrix

In Section 1.1, the relationship between a graph and its corresponding adjacency matrix was discussed. In this section, we will explore some ways to use the adjacency matrix to make further conclusions about data.

Powers of the adjacency matrix can give insight into indirect relationships. The $(i,j)^{\text{th}}$ entry of A^k, where k is a positive integer, represents the number of paths of length exactly k between vertices i and j. Thus the $(i,j)^{\text{th}}$ entry of

$$\sum_{k=0}^{n} A^k$$

represents the number of paths of length n or less between vertices i and j.

Example 1.2. *In the crime network problem presented in Figure 1.8, notice that although Suspect X and Y do share one relationship, M_5, law enforcement may want to dig deeper to see who else is closely related with each of these two* *suspects. Using the adjacency matrix, A, corresponding to this crime network,*

$$
A^2 = \begin{array}{c} \\ M_1 \\ M_2 \\ M_3 \\ M_4 \\ M_5 \\ M_6 \\ M_7 \\ M_8 \\ X \\ Y \end{array}
\begin{array}{c}
\begin{array}{cccccccccc} M_1 & M_2 & M_3 & M_4 & M_5 & M_6 & M_7 & M_8 & X & Y \end{array} \\
\left(\begin{array}{cccccccccc}
4 & 1 & 2 & 1 & 2 & 1 & 2 & 1 & 1 & 1 \\
1 & 2 & 1 & 1 & 0 & 0 & 1 & 1 & 1 & 1 \\
2 & 1 & 3 & 1 & 1 & 1 & 2 & 1 & 1 & 2 \\
1 & 1 & 1 & 5 & 1 & 1 & 2 & 3 & 1 & 1 \\
2 & 0 & 1 & 1 & 4 & 2 & 2 & 2 & 1 & 1 \\
1 & 0 & 1 & 1 & 2 & 3 & 2 & 1 & 2 & 1 \\
2 & 1 & 2 & 2 & 2 & 2 & 4 & 1 & 1 & 1 \\
1 & 1 & 1 & 3 & 2 & 1 & 1 & 6 & 2 & 3 \\
1 & 1 & 1 & 1 & 1 & 2 & 1 & 2 & 3 & 1 \\
1 & 1 & 2 & 1 & 1 & 1 & 1 & 3 & 1 & 4
\end{array}\right)
\end{array} \; and
$$

$$
I + A + A^2 = \begin{array}{c} \\ M_1 \\ M_2 \\ M_3 \\ M_4 \\ M_5 \\ M_6 \\ M_7 \\ M_8 \\ X \\ Y \end{array}
\begin{array}{c}
\begin{array}{cccccccccc} M_1 & M_2 & M_3 & M_4 & M_5 & M_6 & M_7 & M_8 & X & Y \end{array} \\
\left(\begin{array}{cccccccccc}
5 & 2 & 2 & 2 & 2 & 1 & 2 & 2 & 2 & 1 \\
2 & 3 & 1 & 2 & 0 & 0 & 1 & 1 & 1 & 1 \\
2 & 1 & 4 & 2 & 1 & 1 & 3 & 2 & 1 & 2 \\
2 & 2 & 2 & 6 & 1 & 1 & 3 & 3 & 1 & 2 \\
2 & 0 & 1 & 1 & 5 & 3 & 2 & 3 & 2 & 2 \\
1 & 0 & 1 & 1 & 3 & 4 & 2 & 2 & 2 & 2 \\
2 & 1 & 3 & 3 & 2 & 2 & 5 & 2 & 1 & 2 \\
2 & 1 & 2 & 3 & 3 & 2 & 2 & 7 & 3 & 3 \\
2 & 1 & 1 & 1 & 2 & 2 & 1 & 3 & 4 & 1 \\
1 & 1 & 2 & 2 & 2 & 2 & 2 & 3 & 1 & 5
\end{array}\right)
\end{array} .
$$

show that Suspects X and Y, corresponding to rows 9 and 10, are connected to all members in the network through a one or two step path. Focusing in on the rows of these matrices, and thus each member's relationships

$$
\sum_{j=1}^{10} A_{i,j}^2 = \{16, 9, 15, 17, 16, 14, 18, \mathbf{21}, 14, 16\}
$$

$$
\sum_{j=1}^{10} (I + A + A^2)_{i,j} = \{21, 12, 19, 23, 21, 18, 23, \mathbf{28}, 18, 21\}
$$

highlights the fact that M_8 may be the most connected member of network and could be someone that law enforcement investigates further.

Using similar techniques to explore the SEC-East example from Section 1.1, let's say that you wish to rank the SEC-East teams based on their record half way through the 2019 season. How would you determine the #1 football team in the SEC-East prior to the end of the regular season?

Example 1.3. *Let matrix A_{SEC}, defined in Section 1.1, model the wins and loses for each team in the season; however, you may want to incorporate indirect relationships between teams.*

The entries of A^2_{SEC} represent the two step win-lose relationships between teams in the SEC-East.

For example, we know from A_{SEC} that Florida beat South Carolina and that South Carolina beat Georgia, so we might predict that Florida would beat Georgia if they played head-to-head. We see this represented with a value of 1 in the (1,4) entry of A^2_{SEC}.

Similarly, the (4,5) entry in A^2_{SEC} represent two ways that we can think about Georgia dominating over Missouri, through Kentucky or through Vanderbilt.

The matrix $A_{SEC} + A^2_{SEC}$ then shows the number of one step and two step domination between teams.

Using this matrix, we can see that although South Carolina may not have risen to the top of the ranking using just A_{SEC}, they have almost the same number of one- and two-step dominance paths as Florida and Georgia and could be a contender for the number one ranking at the end of the season.

$$A^2_{SEC} = \begin{pmatrix} 0 & 0 & 1 & 1 & 1 & 1 & 0 \\ 0 & 0 & 1 & 1 & 0 & 0 & 0 \\ 0 & 0 & 0 & 0 & 0 & 1 & 0 \\ 0 & 0 & 0 & 0 & 2 & 1 & 0 \\ 0 & 0 & 1 & 1 & 0 & 0 & 0 \\ 0 & 1 & 1 & 0 & 1 & 0 & 1 \\ 0 & 0 & 0 & 0 & 0 & 1 & 0 \end{pmatrix} \quad and$$

$$A_{SEC} + A^2_{SEC} = \begin{pmatrix} 0 & 1 & 2 & 1 & 1 & 2 & 0 \\ 0 & 0 & 1 & 1 & 0 & 1 & 0 \\ 0 & 0 & 0 & 0 & 1 & 1 & 0 \\ 0 & 1 & 1 & 0 & 2 & 1 & 1 \\ 0 & 0 & 1 & 1 & 0 & 1 & 0 \\ 0 & 1 & 2 & 1 & 1 & 0 & 1 \\ 0 & 0 & 0 & 0 & 1 & 1 & 0 \end{pmatrix}.$$

We will look further at techniques for sports ranking in Section 1.6 that not only incorporate win-lose but also the score differential of games.

1.3 Eigenvalues and Eigenvectors as Key Players

If A is a $n \times n$ square matrix and

$$Ax = \lambda x,$$

λ is called an **eigenvalue** of A with corresponding **eigenvector** x, where x is an $n \times 1$ vector.

Given a square matrix A, one way to find the eigenvalues for A is to find the **characteristic equation**

$$|A - \lambda I| = 0$$

and to solve for all values of λ.

The **spectral radius** of a matrix A,

$$\rho(A) = max_{1 \leq i \leq n}|\lambda_i|,$$

where $\lambda_i, 1 \leq i \leq n$, are the eigenvalues of A.

Example 1.4. *If* $A = \begin{pmatrix} 1 & 2 \\ 2 & 4 \end{pmatrix}$ *then to find the eigenvalues of A*

$$|A - \lambda I| = 0$$

$$\begin{vmatrix} 1 - \lambda & 2 \\ 2 & 4 - \lambda \end{vmatrix} = 0$$

$$\lambda^2 - 5\lambda = 0$$

$$\lambda(\lambda - 5) = 0.$$

Thus the eigenvalues of A are $\lambda = 0$ *and* $\lambda = 5$. *The spectral radius*

$$\rho(A) = max\{\ |0|,\ |5|\ \} = 5.$$

Each of these eigenvalues has a corresponding eigenvector, also called a basis vector for the eigenspace. For example, for $\lambda = 5$,

$$\begin{pmatrix} 1 & 2 \\ 2 & 4 \end{pmatrix} \begin{pmatrix} x \\ y \end{pmatrix} = \begin{pmatrix} 5x \\ 5y \end{pmatrix}.$$

Here, $x + 2y = 5x$ *and* $2x + 4y = 5y$. *Therefore,* $x = 0$ *and a basis for the eigenspace corresponding to* $\lambda = 5$ *is* $\{1, 2\}$.

Similarly, when $\lambda = 0$, $x + 2y = 0$ *and* $\{-2, -1\}$ *is a basis vector for the eigenspace corresponding to* $\lambda = 0$.

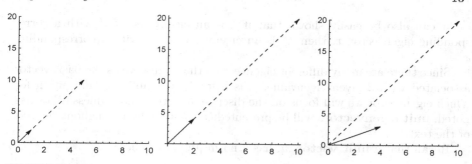

FIGURE 1.10

Av where $A = \begin{pmatrix} 1 & 2 \\ 2 & 4 \end{pmatrix}$, $v = \{1,2\}, \{2,4\}$, and $\{4,3\}$.

In general, a $n \times n$ matrix A acts as a transformation, or function, on the vectors $v \in \mathbb{R}^n$.

Example 1.5. *For example, if $v = \{4, 3\}$ and*

$$A = \begin{pmatrix} 1 & 2 \\ 2 & 4 \end{pmatrix}$$

$$Av = \begin{pmatrix} 1 & 2 \\ 2 & 4 \end{pmatrix} \begin{pmatrix} 4 \\ 3 \end{pmatrix} = \begin{pmatrix} 10 \\ 20 \end{pmatrix}$$

Notice in Figure 1.10 that the matrix A stretches the vectors $\{1, 2\}$ and $\{2, 4\}$ by a factor of 5. In fact, if a vector is an eigenvector associated with a matrix, then the matrix will stretch the eigenvector by a factor equal to its corresponding eigenvalue. The vector $x = \{1, 2\}$ is an eigenvector of A with corresponding eigenvalue $\lambda = 5$. That is,

$$\begin{pmatrix} 1 & 2 \\ 2 & 4 \end{pmatrix} \begin{pmatrix} 1 \\ 2 \end{pmatrix} = 5 \begin{pmatrix} 1 \\ 2 \end{pmatrix}.$$

So, what about the vector $\{2, 4\}$?

It is important to note that there is actually a family of eigenvectors associated with any given eigenvalue. All eigenvectors associated with a given eigenvalue are scalar multiples of one another. We see $\{2, 4\} = 2\{1, 2\}$ and any scalar multiple of the vector $\{1, 2\}$ is an eigenvector corresponding with the eigenvalue $\lambda = 5$ in this example. Hence, any single one of these vectors may be referred to as the basis vector for the eigenspace, or simply the eigenvector.

One of the most significant properties related to eigenvalues is that a matrix is singular if and only if at least of the eigenvalues of the matrix is equal to 0.

It can also be easily shown that if λ is an eigenvalue of A with a corresponding eigenvector x, then λ^k is an eigenvalue of A^k with a corresponding x.

Since there are many different eigenvectors that can serve as the basis vector associated with a given eigenvalue, it is helpful to establish a convention for which eigenvector we will focus on the discussion in this text. Unless otherwise noted, **unit eigenvectors** will be presented conventionally throughout the rest of the text.

Recall that a **unit vector**, u, has a norm $||u|| = 1$, where

$$||u|| = \sqrt{< u,u >}$$

and given a vector v,

$$\frac{v}{||v||}$$

is a unit vector.

The unit eigenvector of A corresponding to $\lambda = 5$ is $\{\frac{1}{\sqrt{5}}, \frac{2}{\sqrt{5}}\}$.

We will apply the concepts of eigenvalues and eigenvectors to another matrix that is important in data science, the **Laplacian matrix**. The Laplacian matrix, L, affiliated with an $n \times n$ adjacency matrix, A, is defined such that

$$L_{ij} = \begin{cases} d_i & \text{if } i = j, \\ -A_{ij} & \text{if } i \neq j, \end{cases}$$

where d_i is the degree of vertex i.

The **normalized Laplacian matrix**, $\mathscr{L}$, is defined as $\mathscr{L} = TLT$ such that T is a diagonal matrix with $T_{ii} = \dfrac{1}{\sqrt{d_i}}$.

Example 1.6. *Let* $A = \begin{pmatrix} 0 & 1 & 1 & 0 \\ 1 & 0 & 1 & 0 \\ 1 & 1 & 0 & 1 \\ 0 & 0 & 1 & 0 \end{pmatrix}$ *, then*

$$L = \begin{pmatrix} 2 & -1 & -1 & 0 \\ -1 & 2 & -1 & 0 \\ -1 & -1 & 3 & -1 \\ -1 & 0 & -1 & 1 \end{pmatrix},$$

The degree of vertex i, d_i, can be found at $L_{i,i}$, so for example, $d_1 = 2$. If one wishes to normalize the Laplacian, L,

$$T = \begin{pmatrix} \frac{1}{\sqrt{2}} & 0 & 0 & 0 \\ 0 & \frac{1}{\sqrt{2}} & 0 & 0 \\ 0 & 0 & \frac{1}{\sqrt{3}} & 0 \\ 0 & 0 & 0 & 1 \end{pmatrix}$$

$$\mathscr{L} = TLT = \begin{pmatrix} 1 & -\frac{1}{2} & -\frac{1}{\sqrt{6}} & 0 \\ -\frac{1}{2} & 1 & -\frac{1}{\sqrt{6}} & 0 \\ -\frac{1}{\sqrt{6}} & -\frac{1}{\sqrt{6}} & 1 & -\frac{1}{\sqrt{3}} \\ 0 & 0 & -\frac{1}{\sqrt{3}} & 1 \end{pmatrix}.$$

$\rho(L) \approx 4$ and $\rho(\mathscr{L}) = 1.72871$.

Example 1.7. *For the complete graph with five vertices, K_5, seen in Figure 1.11, the Laplacian matrix*

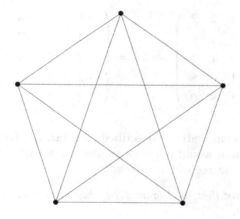

FIGURE 1.11
Complete Graph with five vertices.

$$L = \begin{pmatrix} 4 & -1 & -1 & -1 & -1 \\ -1 & 4 & -1 & -1 & -1 \\ -1 & -1 & 4 & -1 & -1 \\ -1 & -1 & -1 & 4 & -1 \\ -1 & -1 & -1 & -1 & 4 \end{pmatrix}$$

with eigenvalues 5, multiplicity 4, and 0. The normalized Laplacian matrix

$$\mathscr{L} = \begin{pmatrix} 1 & -\frac{1}{4} & -\frac{1}{4} & -\frac{1}{4} & -\frac{1}{4} \\ -\frac{1}{4} & 1 & -\frac{1}{4} & -\frac{1}{4} & -\frac{1}{4} \\ -\frac{1}{4} & -\frac{1}{4} & 1 & -\frac{1}{4} & -\frac{1}{4} \\ -\frac{1}{4} & -\frac{1}{4} & -\frac{1}{4} & 1 & -\frac{1}{4} \\ -\frac{1}{4} & -\frac{1}{4} & -\frac{1}{4} & -\frac{1}{4} & 1 \end{pmatrix}.$$

The Laplacian matrix for simple weighted graphs is very similar to unweighted graphs. In a weighted graph, the entries of the adjacency matrix

$$A_{ij} = \begin{cases} w_{ij} & \text{when vertex } i \text{ and } j \text{ are adjacent} \\ 0 & \text{otherwise} \end{cases}$$

where w_{ij} is the weight of the edge that is incident to vertices i and j. The degree of vertex i, $d_i = \sum w_{ij}$.

Example 1.8. *For the graph, G, pictured in Figure 1.12, the adjacency matrix*

$$A = \begin{pmatrix} 0 & 4 & 4 & 2 & 2 \\ 4 & 0 & 7 & 4 & 4 \\ 4 & 7 & 0 & 1 & 1 \\ 2 & 4 & 1 & 0 & 8 \\ 2 & 4 & 1 & 8 & 0 \end{pmatrix} \text{ and } L = \begin{pmatrix} 12 & -4 & -4 & -2 & -2 \\ -4 & 19 & -7 & -4 & -4 \\ -4 & -7 & 13 & -1 & -1 \\ -2 & -4 & -1 & 15 & -8 \\ -2 & -4 & -1 & -8 & 15 \end{pmatrix}.$$

Unlike the Laplacian matrices described thus far, the Laplacian matrix for a simple directed graph would not be a symmetric matrix. With this in mind, d_i is equal to the outdegree of the vertex.

Example 1.9. *Notice that, in Figure 1.13, that vertex 4 has an outdegree of 3 and an indegree of 1.*

Thus the Laplacian Matrix

$$L = \begin{pmatrix} 2 & -1 & -1 & 0 & 0 \\ 0 & 2 & -1 & 0 & -1 \\ 0 & 0 & 1 & -1 & 0 \\ -1 & -1 & 0 & 3 & -1 \\ -1 & 0 & -1 & 0 & 2 \end{pmatrix}.$$

The Laplacian matrix has significant applications in **spectral graph theory**.

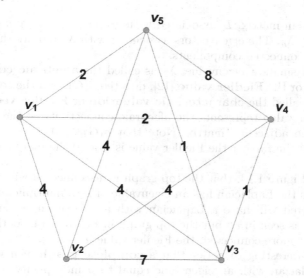

FIGURE 1.12
Example of a weighted K_5 graph.

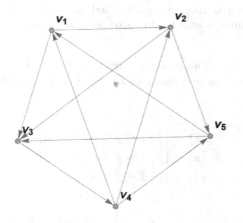

FIGURE 1.13
Example of a directed K_5 graph.

Spectral graph theory is related to clustering and ranking of vertices based on the study of eigenvalues and eigenvectors of the Laplacian matrix. We will focus our attention here on simple connected undirected graphs and their symmetric adjacency matrix counterparts.

The Laplacian matrix, L, associated with graph G will have n eigenvalues, $\lambda_1 \leq \lambda_2 \leq \ldots \leq \lambda_n$. The eigenvectors associated with $\lambda_i = 0$ are the indicators of the graph's connected components.

The second smallest eigenvalue, λ_2, is called the **algebraic connectivity of the graph** or the **Fiedler value** [12], denoted $a(G)$, and the corresponding eigenvector is called the **characteristic valuation** or **Fiedler vector**.

The Fiedler value represents the algebraic connectivity of the graph associated with the adjacency matrix. Note that $a(G) > 0$ if and only if G is connected, and the further the Fiedler value is from 0 the more connected the graph is.

Notice, in Figure 1.14, that the top graph is disconnected with two components and thus the Laplacian has an eigenvalue of zero multiplicity 2. A graph that is connected will have a Laplacian with an eigenvalue equal to zero of multiplicity 1, as seen in all but the top graph in Figure 1.14. As the graphs in Figure 1.14 get more connected, the Fiedler value increases.

It is also interesting to notice that a **complete graph** of n vertices, K_n, will have Laplacian with an eigenvalue equal to n multiplicity $n - 1$ and an eigenvalue equal to 0 multiplicity 1. The Fiedler value and vector of the normalized Laplacian matrix can be interpreted in a similar manner to that of the Laplacian matrix.

If one wishes to partition an undirected graph into two subgraphs, the Fiedler theorem defines a way to do so while minimizing the number of edges cut when doing so.

Theorem 2. *(Fiedler's Theorem for Graph Partitions [Slininger, 2013]) If G is an undirected graph on n vertices and x the Fiedler vector. Define*

$$i_0(x) = \{j \mid 1 \leq j \leq n,\ x_j = 0\}$$
$$i_-(x) = \{j \mid 1 \leq j \leq n,\ x_j < 0\}$$
$$i_+(x) = \{j \mid 1 \leq j \leq n,\ x_j > 0\}.$$

Partition G into $G_1 = i_-(x)$ and $G_2 = i_0(x) \cup i_+(x)$. Under this partitioning, G_1 and G_2 will be connected subgraphs of G.

The ith value in the Fiedler vector is called the **valuation** of vertex i in graph G.

Example 1.10. *The Laplacian matrix corresponding to the graph featured in Figure 1.15*

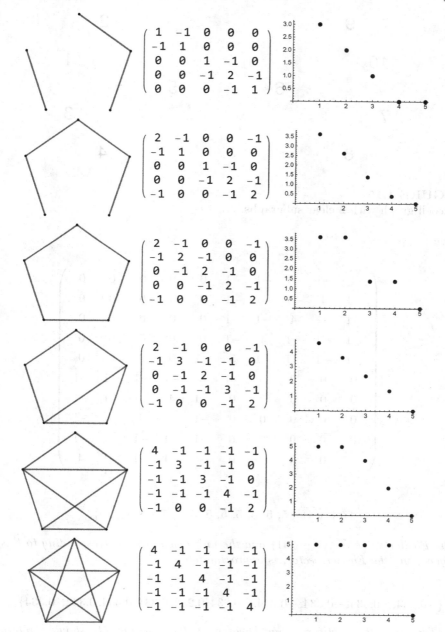

FIGURE 1.14
Examples of Graphs, their Laplacian, and Eigenvalues.

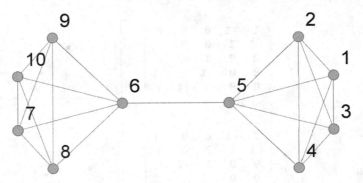

FIGURE 1.15
Barbell graph with 2 clear subgraphs.

$$
L = \begin{pmatrix}
4 & -1 & -1 & -1 & -1 & 0 & 0 & 0 & 0 & 0 \\
-1 & 4 & -1 & -1 & -1 & 0 & 0 & 0 & 0 & 0 \\
-1 & -1 & 4 & -1 & -1 & 0 & 0 & 0 & 0 & 0 \\
-1 & -1 & -1 & 4 & -1 & 0 & 0 & 0 & 0 & 0 \\
-1 & -1 & -1 & -1 & 5 & -1 & 0 & 0 & 0 & 0 \\
0 & 0 & 0 & 0 & -1 & 5 & -1 & -1 & -1 & -1 \\
0 & 0 & 0 & 0 & 0 & -1 & 4 & -1 & -1 & -1 \\
0 & 0 & 0 & 0 & 0 & -1 & -1 & 4 & -1 & -1 \\
0 & 0 & 0 & 0 & 0 & -1 & -1 & -1 & 4 & -1 \\
0 & 0 & 0 & 0 & 0 & -1 & -1 & -1 & -1 & 4
\end{pmatrix}
$$

are

$$\{\tfrac{1}{2}(7 + \sqrt{41}), 5, 5, 5, 5, 5, 5, 5, \tfrac{1}{2}(7 - \sqrt{41}), 0\}.$$

The Fiedler value is $\frac{1}{2}(7 - \sqrt{41})$ and the unit eigenvector corresponding to this eigenvalue, the Fiedler vector, is approximately

$$\{-0.334, -0.334, -0.334, -0.334, -0.234, 0.234, 0.334, 0.334, 0.334, \ 0.334\}.$$

Following the partition from Theorem 2, the subgraph G_1 should contain vertices 1 through 5, while the subgraph G_2 should contain vertices 6 through 10. As can be seen in Figure 1.15, this result is consistent with the clear partitioning in the barbell graph, cutting through just one edge in order to create this partition.

In application, the Fiedler vector is particularly helpful in clustering data that is related to a weighted simple graph. Note that the adjacency matrix must be symmetric in order to apply this partitioning.

There are spectral clustering techniques that can be applied to non-symmetric Laplacian matrices such as that corresponding to the directed graph in Example 1.9. These techniques will be explored further in Chapter 2.

One technique is to use a symmetric matrix, such as the **Gramian matrix** $S = AA^T$ or $A + A^T$, instead of the original adjacency matrix and the Laplacian matrix affiliated with that symmetric matrix.

Example 1.11. *(Crime Network Revisited) Reinvestigating the relationships in the crime network from Figure 1.8, if we wish to cluster the members of this network into two separate clusters, we can do so by using the unit Fiedler vector*

$$\{0.083, 0.179, 0.141, 0.685, -0.262, -0.155, 0.253, -0.474, -0.198, \ 0.228\}.$$

Recall that the Fiedler vector is the eigenvector associated with the 2nd smallest eigenvalue, so when you calculate these, you may get a scalar multiple of the above vector. This should not affect how you cluster the entries.

With a division between those members with positive values in the Fiedler vectors and those with negative values, the two clusters of vertices can be found

$$\{M_1, M_2, M_3, M_4, M_7, SuspectY\} \text{ and } \{SuspectX, M_5, M_6, M_8\}.$$

We can use similar techniques when trying to cluster weighted and directed graphs.

Example 1.12. *Referring back to Example 1.8, the unit Fiedler vector associated with the normalized Laplacian matrix*

$$\mathcal{L} = \begin{pmatrix} 1 & -\frac{2}{\sqrt{57}} & -\frac{2}{\sqrt{39}} & -\frac{1}{3\sqrt{5}} & -\frac{1}{3\sqrt{5}} \\ -\frac{2}{\sqrt{57}} & 1 & -\frac{7}{\sqrt{247}} & -\frac{4}{\sqrt{285}} & -\frac{4}{\sqrt{285}} \\ -\frac{2}{\sqrt{39}} & -\frac{7}{\sqrt{247}} & 1 & -\frac{1}{\sqrt{195}} & -\frac{1}{\sqrt{195}} \\ -\frac{1}{3\sqrt{5}} & -\frac{4}{\sqrt{285}} & -\frac{1}{\sqrt{195}} & 1 & -\frac{8}{15} \\ -\frac{1}{3\sqrt{5}} & -\frac{4}{\sqrt{285}} & -\frac{1}{\sqrt{195}} & -\frac{8}{15} & 1 \end{pmatrix}$$

is

$$\{-0.308, -0.257, -0.529, 0.529, 0.529\}$$

thus clustering together vectors v_1, v_2, and v_3.

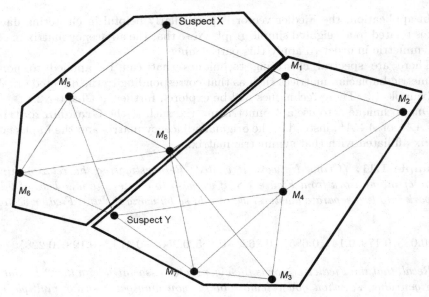

FIGURE 1.16
Crime Network with 2 clear subgraphs.

So why does the Fiedler vector work in clustering data into two subgroups?
We take a moment here to explore the theory behind this concept.

A **quadratic form** of a symmetric matrix A is a function on $\mathbb{R}^n$ where
$Q_A(x) = x^T A x$, or

$$Q_A(x_1, x_2, x_3, \ldots, x_n) = \sum_{i \leq j} a_{ij} x_i x_j.$$

Example 1.13. *Let* $A = \begin{pmatrix} 3 & 2 & -1 \\ 2 & 3 & -1 \\ -1 & -1 & 4 \end{pmatrix}$.

$$Q_A(x) = \begin{pmatrix} x_1 & x_2 & x_3 \end{pmatrix} \cdot \begin{pmatrix} 3 & 2 & -1 \\ 2 & 3 & -1 \\ -1 & -1 & 4 \end{pmatrix} \cdot \begin{pmatrix} x_1 \\ x_2 \\ x_3 \end{pmatrix}$$

$$= 3x_1^2 + 4x_1 x_2 + 3x_2^2 - 2x_1 x_3 - 2x_2 x_3 + 4x_3^2.$$

A real symmetric $n \times n$ matrix, A is called

1. **Positive definite** if $x^T A x > 0$ for all $x \in \mathbb{R}^n$,
2. **Positive semidefinite** if $x^T A x \geq 0$ for all $x \in \mathbb{R}^n$,
3. **Negative definite** if $x^T A x < 0$ for all $x \in \mathbb{R}^n$,
4. **Negative semidefinite** if $x^T A x \leq 0$ for all $x \in \mathbb{R}^n$.

Example 1.14. *Let* $L = \begin{pmatrix} 2 & -1 & 0 & 0 & -1 \\ -1 & 3 & -1 & -1 & 0 \\ 0 & -1 & 2 & -1 & 0 \\ 0 & -1 & -1 & 3 & -1 \\ -1 & 0 & 0 & -1 & 2 \end{pmatrix}$,

$$x^T L x = (x_1 - x_2)^2 + (x_1 - x_5)^2 + (x_2 - x_3)^2 + (x_2 - x_4)^2 + (x_3 - x_4)^2 + (x_4 - x_5)^2 \geq 0$$

and thus L *is a positive semidefinite matrix.*

Lemma 1. *Let A be a $n \times n$ real positive semidefinite matrix, then all of the eigenvalues of A are non-negative.*

Proof. If λ is an eigenvalue of A then there is a vector x such that $Ax = \lambda x$. Thus $0 \leq x^T A x = \lambda x^T x$ and since $x^T x \geq 0$ for all vectors x, $\lambda \geq 0$. $\square$

The Laplacian matrix, L, associated with a non-negative (real) symmetric matrix is a real positive semidefinite matrix and thus has all non-negative eigenvalues. In addition, $L.e = 0$, where $e = \vec{1}$ and since G is connected, e is the only solution to $x^T L x = 0$.

Under the fixed point characterization of eigenvalues, $Ax = \lambda x$, if $|\lambda| < 1$,

$$\lim_{k \to \infty} A^k x = \lim_{k \to \infty} \lambda^k x = \vec{0}.$$

In application, $\lambda = 1$ is of particular interest as $Ax = x$ the values of the corresponding eigenvector, x, remain fixed when multiplied by the matrix A. Similarly, if $|\lambda| = 1$, there exists a positive integer k such that

$$A^k x = x$$

and thus the values of x remain fixed when multiplied by the matrix A^k. We call the eigenvector, x, the **steady state vector**.

Spectral clustering is dependent on the **Spectral Theorem** and another characterization of eigenvalues, the optimization characterization.

Theorem 3. *(Spectral Theorem) Let A be a $n \times n$ real symmetric matrix with eigenvalues $\lambda_1, \ldots, \lambda_n$, then A can be written as*

$$A = U \Lambda U^T,$$

where Λ is a diagonal matrix with i^{th} diagonal entry λ_i and the columns of U, $x_1, \ldots, x_n$, are orthonormal vectors, which are orthonormal bases vectors for the corresponding eigenspaces of A.

The optimization characterization of eigenvalues looks at the **Rayleigh Quotient**

$$R(x) = \frac{x^T A x}{x^T x}.$$

Theorem 4. *(Rayleigh-Ritz Theorem) Let A be a $n \times n$ real symmetric matrix with eigenvalues $\lambda_1 \leq \lambda_2 \leq \cdots \leq \lambda_n$ then*

$$\lambda_1 \leq R(x) \leq \lambda_n.$$

Proof. Since A is a symmetric matrix with eigenvalues $\lambda_1 \leq \lambda_2 \leq \cdots \leq \lambda_n$, from the Spectral Theorem, there is an orthonormal set, $\{x_1, \ldots, x_n\}$, such that for every eigenvector x, $x = \sum_i k_i x_i$, where k_i are scalars.

$$
\begin{aligned}
R(x) &= \frac{(\sum_i k_i x_i)^T A (\sum_i k_i x_i)}{(\sum_i k_i x_i)^T (\sum_i k_i x_i)} \\
&= \frac{(\sum_i k_i x_i)^T A (\sum_i k_i x_i)}{<\sum_i k_i x_i, k_i x_i>} \\
&= \frac{\sum_i k_i^2 \lambda_i}{\sum_i k_i^2}.
\end{aligned}
$$

This shows that $R(x)$ is a weighted sum of the eigenvalues and thus

$$\lambda_1 \leq R(x) \leq \lambda_n.$$

$\square$

From Theorem 7.8, we can conclude that λ_1 minimizes $R(x)$ and λ_n maximizes $R(x)$.

For the $n \times n$ Laplacian matrix, L, of a connected graph, G, with eigenvalues $0 = \lambda_1 < \lambda_2 \leq \cdots \leq \lambda_n$,

$$\lambda_2 = a(G) = min_{x \neq 0, ||x||=1, x \perp e} \frac{x^T L x}{x^T x}.$$

Now if the main goal is to partition the vertices in a graph, G, into two subsets, G_1 and G_2, one might wish to do this without cutting through a large number of edges of the graph. We will measure how well we do this by defining the subset of vertices V_1 and the boundary of V_1,

$$\partial(V_1) = \{(v_1, v_2), v_1 \in V_1, v_2 \in V \setminus V_2\}.$$

The **isoperimetric ratio** [2,3] of V_1, $\theta(V_1)$, is defined as

$$\theta(V_1) = \frac{|\partial(V_1)|}{|V_1|}.$$

The **isoperimetric number** of a graph, θ_G, is the minimum isoperimetric ratio over all sets of at most half of the vertices. That is

$$\theta_G = \min_{|V_1| \leq \frac{|V|}{2}} \frac{|\partial(V_1)|}{|V_1||V_2|}.$$

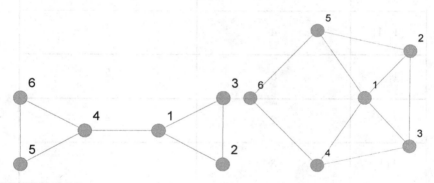

FIGURE 1.17
Graphs with minimal isoperimetric ratio determined with one edge cut (left) and three edges cut (right).

Example 1.15. *Referring to Figure 1.17 (left), $|V| = 6$ and thus we wish to find a subset, V_1, such that $|V_1| \leq 3$ which minimizes θ_G. Table 1.1 shows that setting $V_1 = \{1, 2, 3\}$ and $V_2 = \{4, 5, 6\}$ minimizes θ_G and only one edge was cut in doing so.*

In the graph pictured in Figure 1.17 (right), the partition $V_1 = \{5, 6\}$ and $V_2 = \{1, 2, 3, 4\}$ minimizes isoperimetric ratio and only cuts through three edges.

TABLE 1.1

Sample of subgraphs and isoperimetric numbers for Figure 1.17 (left).

V_1	Number of Cut Edges	Isoperimetric Number, θ_G
$\{1, 2, 3\}$	1	$\frac{1}{9}$
$\{1, 2, 4\}$	4	$\frac{4}{9}$
$\{1, 2, 5\}$	5	$\frac{5}{9}$
$\{2, 3\}$	2	$\frac{2}{8}$
$\{1, 2\}$	3	$\frac{3}{8}$
$\{1, 4\}$	4	$\frac{4}{8}$
$\{1, 5\}$	5	$\frac{5}{8}$
$\{2\}$	2	$\frac{2}{5}$
$\{1\}$	3	$\frac{3}{5}$

1.4   CASE STUDY: Applications in Sport Ranking

There are several ways to incorporate the score differential of games into a sports ranking. We explore a few of these here.

In Section 1.2, we only incorporated a win-lose structure in the analysis of the SEC-East Conference ranking. Here we wish to rank the teams by incorporating sequences of both wins and losses. These paths could arbitrarily include a game where a team plays themselves, represented by a loop in a graph or one in a main diagonal entry of the adjacency matrix. For further analysis, we will use the adjacency matrix

$$A_{SEC*} = A_{SEC} + A_{SEC}^T + I.$$

We begin this exploration of sports ranking by introducing the vector $\vec{S}$, representing the net score for each team, where

$$\vec{S}_i = \sum_{j=1}^{n} \frac{s_{ij}}{n_i},$$

n_i represents the number of games played by team i and s_{ij} represents the point differential in the game between team i and team j, noting that if team i loses to team j, then s_{ij} is negative [19]. From Figure 1.18,

$$\vec{S} = \begin{pmatrix} \dfrac{50}{3} \\[2mm] \dfrac{-40}{3} \\[2mm] -6 \\[2mm] \dfrac{70}{4} \\[2mm] -3 \\[2mm] -6 \\[2mm] \dfrac{-17}{2} \end{pmatrix} \begin{matrix} FL \\[2mm] TN \\[2mm] KY \\[2mm] GA \\[2mm] MIZ \\[2mm] SC \\[2mm] VAN \end{matrix}.$$

With the matrices A_{SEC*} and $\vec{S}$, we can represent generational rankings. First generational rankings from head-to-head matches can be expressed by

$$A_{SEC*}^0 \vec{S},$$

ranking Georgia as #1, Florida as #2, and Missouri as #3.

However, we may wish to include 2nd generation games. These are indirect relationships in a series of competitions, or tournaments, connected by one edge.

For example, in Figure 1.18, although Georgia did not play Missouri, we can say that there is a 2nd generation game between these two teams since Georgia played Vanderbilt and Vanderbilt played Missouri. A ranking that includes 1st,

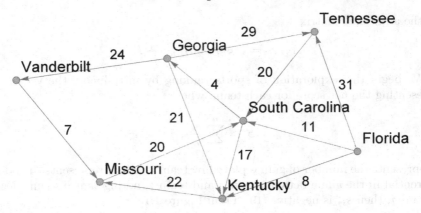

FIGURE 1.18
Weighted Directed Graph for SEC-East Football Example.

2nd, and 3rd generation games can be found with

$$\vec{S} + (A_{SEC*})\vec{S} + (A_{SEC*})^2\vec{S}$$
$$= \{39.2, \ -2.8, \ -10.3, \ 30.7, \ -19, \ -8.8, \ -36.3\}$$

ranking Florida above Georgia, with Tennessee coming in third. Generalizing this idea, a ranking that includes the first k generation games can be found with

$$\sum_{n=0}^{k} A_{SEC*}^n \vec{S}.$$

Another way to rank the teams is to incorporate the points scored for each team, defining

$$A_{ij} = \frac{W_{ij}}{(W_{ij} + W_{ji})}$$

where W_{ij} is the score of team i in the game between teams i and j. The scores of the games are shown in Figure 1.18.

TABLE 1.2
2019 SEC-East Football Scores.

FL(29)-KY(21)	FL(34)-TN(3)	FL(38)-SC(27)
TN(14)-GA(43)	TN(41)-SC(21)	KY(7)-SC(24)
KY(0)-GA(21)	KY(29)-MIZ(7)	GA(30)-VAN(6)
GA(17)-SC(21)	MIZ(34)-SC(14)	MIZ(14)-VAN(21)

The absolute value of the unit eigenvector

$$\{0.434, 0.293, 0.307, 0.514, 0.282, 0.515, 0.149\}$$

corresponding to the largest eigenvalue with a norm equal to 1, in absolute value, associated with this scoring technique provides a ranking of South Carolina, Georgia, and then Florida. This ranking technique incorporates the team's relative score; however, it does not incorporate the notion of a closely fought game, a game won in overtime, or when a win occurs in the season.

In order to avoid having a team run up the score in a single game in order to greatly improve their team rankings, some mathematicians use a slightly different weighting system where

$$A_{ij} = h\frac{W_{ij} + 1}{(W_{ij} + W_{ji} + 2)}$$

and $h \in (0,1]$.

Wesley Colley developed a weighted system that is believed to deal with some of the issues seen in the previous technique. In the Colley Method [6] of sports ranking, the entries of C, the **Colley matrix**, are defined as

$$C_{ij} = \begin{cases} 2 + N_i & when\, i = j \\ -n_{ji} & when\, i \neq j \end{cases},$$

where N_i is the number of games played by team i and n_{ji} is the number of times team i has played team j. The goal of this ranking system is to solve for $\vec{r}$ in the linear system

$$C\vec{r} = b \text{ where } b_i = 1 + \frac{1}{2}(w_i - l_i),$$

w_i is the number of wins for team i and l_i is the number of losses for team i.

For the SEC-East Conference, the Colley Method, defined with the linear system $Cr = b$, with

$$C = \begin{pmatrix} 5 & -1 & -1 & 0 & 0 & -1 & 0 \\ -1 & 5 & 0 & -1 & 0 & -1 & 0 \\ -1 & 0 & 6 & -1 & -1 & -1 & 0 \\ 0 & -1 & -1 & 6 & 0 & -1 & -1 \\ 0 & 0 & -1 & 0 & 5 & -1 & -1 \\ -1 & -1 & -1 & -1 & -1 & 7 & 0 \\ 0 & 0 & 0 & -1 & -1 & 0 & 4 \end{pmatrix}$$

and $b = \{\frac{5}{2}, \frac{1}{2}, 0, 2, \frac{1}{2}, \frac{1}{2}, 1\}$ produces a ranking vector,

$$r = \{0.753, 0.627, 0.497, 0.464, 0.438, 0.363, 0.359\}.$$

Under this Colley method, Florida would be ranked #1 with Georgia and Vanderbilt 2nd and 3rd respectively.

This case study presents several different ways to use matrices to produce a ranking among sports teams. There is no single way to do a sport ranking, and are many other famous ranking systems such as Massey method and Ford's method [6].

Some data used in this study can be found at 3. A larger data set of games from the NCAA Men's Division 1 2020 season can be found at GitHub links 4 and 5. Python and R code for this case study can be found at GitHub links 2 and 7 respectively.

1.5   CASE STUDY: Gerrymandering

Throughout American history, states have struggled with how to create voting districts in a non-partisan manner. As recent as 2019, states, such as North Carolina, have had congressional districts ruled unconstitutional due to partisan gerrymander [39]. The 4th Circuit Court of Appeals has made a clear statement that focuses on three steps to determine if a district has been gerrymandered, violating the Constitution.

A district should be in question if there is "(1) discriminatory intent, (2) discriminatory effects, and (3) a lack of justification for the discriminatory effects" [2].

In order to think mathematically about how to develop fair districts, we must also know some rules around districting. First of all, districts are typically created using defined census areas, census blocks, census block groups, or census tracts. The former is the smallest and the later is the largest of the defined areas. Districts must also maintain three characteristics: equity, contiguity, and compactness [15].

Equity is the principle that all districts have an equivalent number of voters. In an ideal situation, districts are compact if given any two points in the district the line segment between these points remains completely in the district. Keep in mind that census blocks, groups, and tracts are typically not compact and thus the inclusion of these objects, in total, may cause a district to fail this property. Failure to meet the property of compactness can become more of an issue when it occurs in an unnatural manner. Failure to meet compactness in the 2011 North Carolina Congressional redistricting map was voted unconstitutional, particularly around District 12, which can be seen in Figure 1.19.

Some of the issues with districting come down to determining adjacent census blocks. As you can see in Figure 1.20, adjacent census blocks may not have consecutive numbers. For example, census blocks 2003 and 2009 are adjacent.

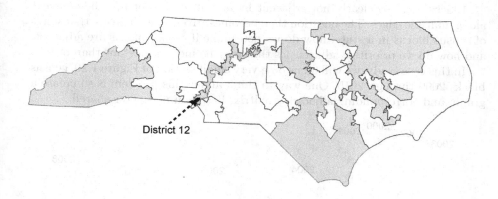

FIGURE 1.19
2011 North Carolina Congressional Redistricting Map [26].

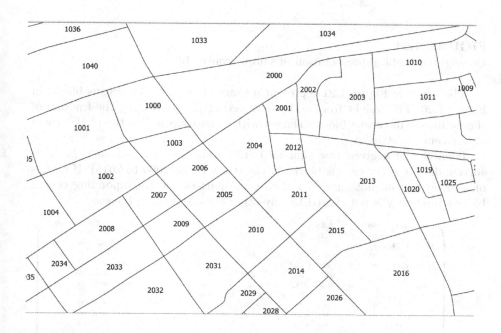

FIGURE 1.20
Sample of census blocks and block number in Charlotte, NC.

Blocks are also clearly not adjacent by an entire edge or may be adjacent along multiple edges. The question then becomes, given that there are thousands of census blocks in a state, how do we determine if census blocks are adjacent, and how do we prioritize adjacencies that may be more important than others.

In this case study, we will focus on a very small section of Figure 1.20, census blocks 2000 through 2015. One way to think about this problem is to create a graph, and corresponding adjacency matrix, based on perimeter shared.

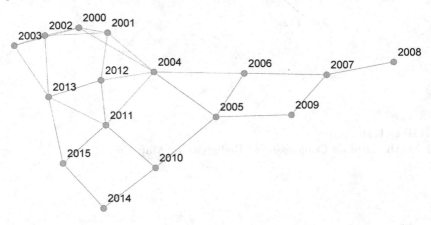

FIGURE 1.21
Unweighted graph representation of sample census blocks.

The edges in Figure 1.21 represent a shared edge between census blocks in Figure 1.20. The weight from vertex i to vertex j, representing the fraction of the boundary of census block i shared with census block j, is the $(i,j)^{\text{th}}$ entry of adjacency matrix, A_E.

If entries of a given row sum to 1, this indicates that a census block is surrounded by other census blocks in the study (from 2000 to 2015). If entries of a given row do not sum to 1 then some part of the corresponding census block's boundary is not shared by any other census block in the study.

$$A_E = \begin{pmatrix}
0 & 0.06 & 0.08 & 0.05 & 0.1 & 0 & 0 & 0 & 0 & 0 & 0 & 0 & 0 & 0 & 0 & 0 \\
0.27 & 0 & 0.34 & 0 & 0.2 & 0 & 0 & 0 & 0 & 0 & 0 & 0 & 0.19 & 0 & 0 & 0 \\
0.24 & 0.24 & 0 & 0.44 & 0 & 0 & 0 & 0 & 0 & 0 & 0 & 0 & 0 & 0.08 & 0 & 0 \\
0.11 & 0 & 0.35 & 0 & 0 & 0 & 0 & 0 & 0 & 0 & 0 & 0 & 0 & 0.2 & 0 & 0 \\
0.28 & 0.13 & 0 & 0 & 0 & 0.14 & 0.15 & 0 & 0 & 0 & 0 & 0.14 & 0.16 & 0 & 0 & 0 \\
0 & 0 & 0 & 0 & 0.19 & 0 & 0.31 & 0 & 0 & 0.19 & 0.31 & 0 & 0 & 0 & 0 & 0 \\
0 & 0 & 0 & 0 & 0.2 & 0.3 & 0 & 0.2 & 0 & 0 & 0 & 0 & 0 & 0 & 0 & 0 \\
0 & 0 & 0 & 0 & 0 & 0 & 0.21 & 0 & 0.21 & 0.58 & 0 & 0 & 0 & 0 & 0 & 0 \\
0 & 0 & 0 & 0 & 0 & 0 & 0 & 0.16 & 0 & 0 & 0 & 0 & 0 & 0 & 0 & 0 \\
0 & 0 & 0 & 0 & 0 & 0.21 & 0 & 0.62 & 0 & 0 & 0 & 0 & 0 & 0 & 0 & 0 \\
0 & 0 & 0 & 0 & 0 & 0.24 & 0 & 0 & 0 & 0 & 0 & 0.24 & 0 & 0 & 0.24 & 0 \\
0 & 0 & 0 & 0 & 0.16 & 0 & 0 & 0 & 0 & 0 & 0.26 & 0 & 0.13 & 0.19 & 0 & 0.26 \\
0 & 0.23 & 0 & 0 & 0.30 & 0 & 0 & 0 & 0 & 0 & 0 & 0.22 & 0 & 0.25 & 0 & 0 \\
0 & 0 & 0.05 & 0.16 & 0 & 0 & 0 & 0 & 0 & 0 & 0 & 0.12 & 0.09 & 0 & 0 & 0.17 \\
0 & 0 & 0 & 0 & 0 & 0 & 0 & 0 & 0 & 0 & 0.25 & 0 & 0 & 0 & 0 & 0.2 \\
0 & 0 & 0 & 0 & 0 & 0 & 0 & 0 & 0 & 0 & 0 & 0.26 & 0 & 0.27 & 0.21 & 0
\end{pmatrix}.$$

In Chapter 1.6, a netscore vector was created in order to rank teams in the SEC-East Conference. A similar vector can be created for studying the census blocks. Let

$$\vec{S}_i = \sum_{j=1}^{16} A_E(i,j) = \{0.29, 1, 1, 0.66, 1, 1, 0.7, 1, 0.16, 0.83, 0.72, 1, 1,$$

$$0.59, 0.45, 0.74\}.$$

$\vec{S}_i$ represents the percent of the boundary of census block i shared by other census blocks in the study. Notice that A_E is not symmetric; however,

$$A_{E*} = A_E + A_E^T + I \text{ is symmetric.}$$

$$\vec{S} + \vec{S}A_{E*} = \{1.7, 3.4, 3.27, 2.37, 3.9, 3.5, 2.8, 3.34, 0.7, 3.26, 2.7, 3.6,$$

$$3.4, 2.5, 1.6, 2.44\}$$

represents a ranking of census blocks where the highest rank blocks have the most boundary shared in the study, census blocks 2004, 2011, 2005, 2012, and 2001.

Unfortunately, the above technique does not provide a clear way to incorporate adjacency into a way to partition the census blocks. We will look to find the Fiedler vector in order to aid us in accomplishing this goal.

If we create a binary adjacency matrix, A, based on whether a census block shares edges with other census blocks in Figure 1.20, the matrix

$$A = \begin{pmatrix}
0 & 1 & 1 & 1 & 1 & 0 & 0 & 0 & 0 & 0 & 0 & 0 & 0 & 0 & 0 & 0 \\
1 & 0 & 1 & 0 & 1 & 0 & 0 & 0 & 0 & 0 & 0 & 0 & 1 & 0 & 0 & 0 \\
1 & 1 & 0 & 1 & 0 & 0 & 0 & 0 & 0 & 0 & 0 & 0 & 0 & 1 & 0 & 0 \\
1 & 0 & 1 & 0 & 0 & 0 & 0 & 0 & 0 & 0 & 0 & 0 & 0 & 1 & 0 & 0 \\
1 & 1 & 0 & 0 & 0 & 1 & 1 & 0 & 0 & 0 & 0 & 1 & 1 & 0 & 0 & 0 \\
0 & 0 & 0 & 0 & 1 & 0 & 1 & 0 & 0 & 1 & 1 & 0 & 0 & 0 & 0 & 0 \\
0 & 0 & 0 & 0 & 1 & 1 & 0 & 1 & 0 & 0 & 0 & 0 & 0 & 0 & 0 & 0 \\
0 & 0 & 0 & 0 & 0 & 0 & 1 & 0 & 1 & 1 & 0 & 0 & 0 & 0 & 0 & 0 \\
0 & 0 & 0 & 0 & 0 & 0 & 0 & 1 & 0 & 0 & 0 & 0 & 0 & 0 & 0 & 0 \\
0 & 0 & 0 & 0 & 0 & 1 & 0 & 1 & 0 & 0 & 0 & 0 & 0 & 0 & 0 & 0 \\
0 & 0 & 0 & 0 & 0 & 1 & 0 & 0 & 0 & 0 & 0 & 1 & 0 & 0 & 1 & 0 \\
0 & 0 & 0 & 0 & 1 & 0 & 0 & 0 & 0 & 0 & 1 & 0 & 1 & 1 & 0 & 1 \\
0 & 1 & 0 & 0 & 1 & 0 & 0 & 0 & 0 & 0 & 0 & 1 & 0 & 1 & 0 & 0 \\
0 & 0 & 1 & 1 & 0 & 0 & 0 & 0 & 0 & 0 & 0 & 1 & 1 & 0 & 0 & 1 \\
0 & 0 & 0 & 0 & 0 & 0 & 0 & 0 & 0 & 0 & 1 & 0 & 0 & 0 & 0 & 1 \\
0 & 0 & 0 & 0 & 0 & 0 & 0 & 0 & 0 & 0 & 0 & 1 & 0 & 1 & 1 & 0 \\
\end{pmatrix}$$

and

$$I + A + A^2 = \begin{pmatrix}
5 & 3 & 3 & 2 & 2 & 1 & 1 & 0 & 0 & 0 & 0 & 1 & 2 & 2 & 0 & 0 \\
3 & 5 & 2 & 2 & 3 & 1 & 1 & 0 & 0 & 0 & 0 & 2 & 2 & 2 & 0 & 0 \\
3 & 2 & 5 & 3 & 2 & 0 & 0 & 0 & 0 & 0 & 0 & 1 & 2 & 2 & 0 & 1 \\
2 & 2 & 3 & 4 & 1 & 0 & 0 & 0 & 0 & 0 & 0 & 1 & 1 & 2 & 0 & 1 \\
2 & 3 & 2 & 1 & 7 & 2 & 2 & 1 & 0 & 1 & 2 & 2 & 3 & 2 & 0 & 1 \\
1 & 1 & 0 & 0 & 2 & 5 & 2 & 2 & 0 & 1 & 1 & 2 & 1 & 0 & 1 & 0 \\
1 & 1 & 0 & 0 & 2 & 2 & 4 & 1 & 1 & 2 & 1 & 1 & 1 & 0 & 0 & 0 \\
0 & 0 & 0 & 0 & 1 & 2 & 1 & 4 & 1 & 1 & 0 & 0 & 0 & 0 & 0 & 0 \\
0 & 0 & 0 & 0 & 0 & 0 & 1 & 1 & 2 & 1 & 0 & 0 & 0 & 0 & 0 & 0 \\
0 & 0 & 0 & 0 & 1 & 1 & 2 & 1 & 1 & 3 & 1 & 0 & 0 & 0 & 0 & 0 \\
0 & 0 & 0 & 0 & 2 & 1 & 1 & 0 & 0 & 1 & 4 & 1 & 1 & 1 & 1 & 2 \\
1 & 2 & 1 & 1 & 2 & 2 & 1 & 0 & 0 & 0 & 1 & 6 & 3 & 3 & 2 & 2 \\
2 & 2 & 2 & 1 & 3 & 1 & 1 & 0 & 0 & 0 & 1 & 3 & 5 & 2 & 0 & 2 \\
2 & 2 & 2 & 2 & 2 & 0 & 0 & 0 & 0 & 0 & 1 & 3 & 2 & 6 & 1 & 2 \\
0 & 0 & 0 & 0 & 0 & 1 & 0 & 0 & 0 & 0 & 1 & 2 & 0 & 1 & 3 & 1 \\
0 & 0 & 1 & 1 & 1 & 0 & 0 & 0 & 0 & 0 & 2 & 2 & 2 & 2 & 1 & 4
\end{pmatrix}$$

show that census blocks 2004 and 2011 have the most blocks adjacent to them in one or two steps, with the total of the rows of

$$I + A + A^2 = \{22, 23, 21, 17, 31, 19, 17, 10, 5, 10, 15, 27, 25, 25, 9, 16\}.$$

One might wish to make sure that these two blocks are in separate districts. However, what if 2004 and 2011 were adjacent to one another?

The Laplacian of A,

$$L = \begin{pmatrix}
4 & -1 & -1 & -1 & -1 & 0 & 0 & 0 & 0 & 0 & 0 & 0 & 0 & 0 & 0 & 0 \\
-1 & 4 & -1 & 0 & -1 & 0 & 0 & 0 & 0 & 0 & 0 & 0 & -1 & 0 & 0 & 0 \\
-1 & -1 & 4 & -1 & 0 & 0 & 0 & 0 & 0 & 0 & 0 & 0 & 0 & -1 & 0 & 0 \\
-1 & 0 & -1 & 3 & 0 & 0 & 0 & 0 & 0 & 0 & 0 & 0 & 0 & -1 & 0 & 0 \\
-1 & -1 & 0 & 0 & 6 & -1 & -1 & 0 & 0 & 0 & 0 & -1 & -1 & 0 & 0 & 0 \\
0 & 0 & 0 & 0 & -1 & 4 & -1 & 0 & 0 & -1 & -1 & 0 & 0 & 0 & 0 & 0 \\
0 & 0 & 0 & 0 & -1 & -1 & 3 & -1 & 0 & 0 & 0 & 0 & 0 & 0 & 0 & 0 \\
0 & 0 & 0 & 0 & 0 & 0 & -1 & 3 & -1 & -1 & 0 & 0 & 0 & 0 & 0 & 0 \\
0 & 0 & 0 & 0 & 0 & 0 & 0 & -1 & 1 & 0 & 0 & 0 & 0 & 0 & 0 & 0 \\
0 & 0 & 0 & 0 & 0 & -1 & 0 & -1 & 0 & 2 & 0 & 0 & 0 & 0 & 0 & 0 \\
0 & 0 & 0 & 0 & 0 & -1 & 0 & 0 & 0 & 0 & 3 & -1 & 0 & 0 & -1 & 0 \\
0 & 0 & 0 & 0 & -1 & 0 & 0 & 0 & 0 & 0 & -1 & 5 & -1 & -1 & 0 & -1 \\
0 & -1 & 0 & 0 & -1 & 0 & 0 & 0 & 0 & 0 & 0 & -1 & 4 & -1 & 0 & 0 \\
0 & 0 & -1 & -1 & 0 & 0 & 0 & 0 & 0 & 0 & 0 & -1 & -1 & 5 & 0 & -1 \\
0 & 0 & 0 & 0 & 0 & 0 & 0 & 0 & 0 & 0 & -1 & 0 & 0 & 0 & 2 & -1 \\
0 & 0 & 0 & 0 & 0 & 0 & 0 & 0 & 0 & 0 & 0 & -1 & 0 & -1 & -1 & 3
\end{pmatrix}$$

has unit Fiedler vector

$$\{0.17, 0.16, 0.2, 0.2, 0.06, -0.1, -0.17, -0.43, -0.64, -0.31, 0.07, 0.14, 0.15, \\ 0.18, 0.15, 0.18\}.$$

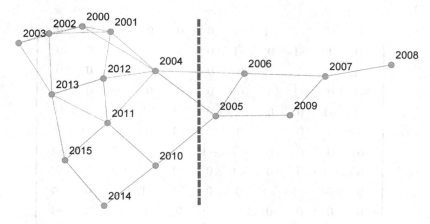

FIGURE 1.22
Graphically representation of sample census blocks and partition.

Based on the Fiedler vector, these census blocks can be partitioned into the two subgraphs, or districts,

$$G_1 = \{2000, 2001, 2002, 2003, 2004, 2010, 2011, 2012, 2013, 2014, 2015, 2016\}$$

$$\text{and } G_2 = \{2005, 2006, 2007, 2008, 2009\}.$$

As described, Fiedler clustering would create two distinct clusters in a graph. However, this process can be repeated in order to create clusters within a subgraph [3]. For example, perhaps it is essential to have three voting districts, let's create a Fiedler vector for G_1, so that we can divide G_1 into two clusters. The adjacency and Laplacian matrix corresponding to G_1 are

$$A_1 = \begin{pmatrix} 0 & 1 & 1 & 1 & 0 & 0 & 0 & 0 & 0 & 0 & 0 \\ 1 & 0 & 1 & 0 & 0 & 0 & 0 & 1 & 0 & 0 & 0 \\ 1 & 1 & 0 & 1 & 0 & 0 & 0 & 0 & 1 & 0 & 0 \\ 1 & 0 & 1 & 0 & 0 & 0 & 0 & 0 & 1 & 0 & 0 \\ 0 & 0 & 0 & 0 & 0 & 0 & 0 & 0 & 0 & 0 & 0 \\ 0 & 0 & 0 & 0 & 0 & 0 & 1 & 0 & 0 & 1 & 0 \\ 0 & 0 & 0 & 0 & 0 & 1 & 0 & 1 & 1 & 0 & 1 \\ 0 & 1 & 0 & 0 & 0 & 0 & 1 & 0 & 1 & 0 & 0 \\ 0 & 0 & 1 & 1 & 0 & 0 & 1 & 1 & 0 & 0 & 1 \\ 0 & 0 & 0 & 0 & 0 & 1 & 0 & 0 & 0 & 0 & 1 \\ 0 & 0 & 0 & 0 & 0 & 0 & 1 & 0 & 1 & 1 & 0 \end{pmatrix} \text{ and }$$

$$L_1 = \begin{pmatrix} 4 & -1 & -1 & -1 & -1 & 0 & 0 & 0 & 0 & 0 & 0 \\ -1 & 4 & -1 & 0 & -1 & 0 & 0 & -1 & 0 & 0 & 0 \\ -1 & -1 & 4 & -1 & 0 & 0 & 0 & 0 & -1 & 0 & 0 \\ -1 & 0 & -1 & 3 & 0 & 0 & 0 & 0 & -1 & 0 & 0 \\ -1 & -1 & 0 & 0 & 4 & 0 & -1 & -1 & 0 & 0 & 0 \\ 0 & 0 & 0 & 0 & 0 & 2 & -1 & 0 & 0 & -1 & 0 \\ 0 & 0 & 0 & 0 & -1 & -1 & 5 & -1 & -1 & 0 & -1 \\ 0 & -1 & 0 & 0 & -1 & 0 & -1 & 4 & -1 & 0 & 0 \\ 0 & 0 & -1 & -1 & 0 & 0 & -1 & -1 & 5 & 0 & -1 \\ 0 & 0 & 0 & 0 & 0 & -1 & 0 & 0 & 0 & 2 & -1 \\ 0 & 0 & 0 & 0 & 0 & 0 & -1 & 0 & -1 & -1 & 3 \end{pmatrix}.$$

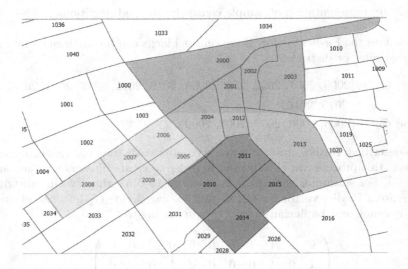

FIGURE 1.23
Subgraphs defined using Fiedler vectors.

Eigenvalues of L_1 are $\{6.6, 6.4, 5.7, 5.2, 4.2, 4., 2.8, 2.3, 2.1, 0.7, 0\}$. The unit Fiedler vector, and thus the unit eigenvector corresponding to eigenvalue 0.7, is

$$\frac{1}{\sqrt{1499}}\{-11, -10, -10, -10, -7, 20, 4, -5, -2, 22, 10\}.$$

Using this information, we can divide G_1 into two districts

$$\{2000, 2001, 2002, 2003, 2004, 2012, 2013\} \text{ and } \{2010, 2011, 2014, 2015\}.$$

The three subgraph of G defined by with the Fiedler vector technique described in this case study can be seen in Figure 1.23.

Data related to the small weighted adjacency matrix for NC census blocks 2000-2015 and full information on the 2010 NC census blocks can be found at Github links 9 and 10. Python and R code for this case study can be found at GitHub links 6 and 8 respectively.

1.6 Exercises

1. Six students from Connected University are trying to figure out how they are connected to one another. They all have some connection to one another through the same online network platform. The adjacency matrix representing their connections to one another is

$$A = \begin{pmatrix} 0 & 1 & 0 & 1 & 0 & 0 \\ 1 & 0 & 1 & 1 & 0 & 1 \\ 0 & 1 & 0 & 0 & 1 & 0 \\ 1 & 1 & 0 & 0 & 1 & 0 \\ 0 & 0 & 1 & 1 & 0 & 1 \\ 0 & 1 & 0 & 0 & 1 & 0 \end{pmatrix}$$

a. Kush is a student in the network whose network connections are represented in Row 1, or Column 1. Using the adjacency matrix, determine how many others are directly connected, through a 1-step path, to Kush in the network.

b. Using the adjacency matrix, determine how many others are directly connected, through a 2-step path, to Kush in the network.

c. Fiona is also a student in the network. Fiona's network connections are represented in Row 2, or Column 2 of the adjacency matrix. Using the adjacency matrix, determine how many 1-step or 2-step connections there between Kush and Fiona.

2. For the matrix

$$A = \begin{pmatrix} 1 & 0 & 0 \\ 2 & 0 & 1 \\ 0 & 1 & 2 \end{pmatrix}$$

find the spectral radius, $\rho(A)$.

3. A graph, G, is associated with the adjacency matrix

$$A = \begin{pmatrix} 1 & 1 & 1 & 0 \\ 1 & 0 & 0 & 1 \\ 1 & 0 & 0 & 1 \\ 0 & 1 & 1 & 0 \end{pmatrix}.$$

Use the adjacency matrix to determine if G a simple connected graph.

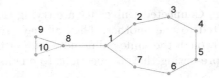

FIGURE 1.24
Figure for Exercises 4 and 5.

4. For the following exercises, refer to graph G in Figure 1.24.

   a.   Find the adjacency matrix corresponding to G.
   b.   Find the Laplacian matrix associated with graph G.
   c.   Calculate the Fiedler value and Fiedler vector.
   d.   Use the Fiedler vector to cluster the graph into two subgraphs.

5. Using graph G displayed in Figure 1.24, calculate the isoperimetric number, θ_G, to argue how to cluster the G into two subgraphs.

6. Wey and Blumstein [37] studied the yellow-bellied marmot, a social ground dwelling rodent, and their network of social interactions in and around the Rocky Mountain Biological Laboratory in Colorado. Assume that the Marmot social network can be represented by the graph G found in Figure 1.25.

   a.   Find the adjacency matrix corresponding to G.
   b.   Let's assume that a marmot's social network can extend through friend marmots, leading us to explore powers of the adjacency matrix from part a. Determine if there is any marmot in the network that is connected to every other marmot through direct or two step relationships. Which marmots in the network are connected to every other marmot through direct, two-step, or three-step relationships?
   c.   Find the Laplacian matrix associated with G.
   d.   Determine the isoperimetric ratio $\theta(V)$ where

$$V = \{v_1,\ v_2,\ v_3,\ v_5,\ v_6,\ v_7\}$$

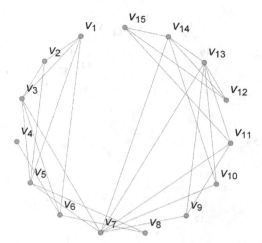

FIGURE 1.25
Marmot social network in Exercise 6.

        affiliated with the clustering the graph G into the two subgraphs G_1 and G_2.

  e. Find the Fiedler vector affiliated with G.

  f. Use the Fiedler vector to cluster the marmot social network represented by G into two clusters.

  g. Use your result in Part e. to cluster the marmot social network into four clusters. There are multiple answers to this question. Argue your solution based on Fiedler vectors of subgraphs of G.

7. If we knew more about the marmot social interactions regarding dominance and number of interactions, how might this affect how the relationships are represented in the graph presented in Figure 1.25.

8. The weighted graph shown in Figure 1.26 displays the number of interactions between corresponding marmots in a week long period. Use this graph to answer the following questions.

  a. Find the Laplacian matrix affiliated with the graph in Figure 1.26.

  b. Find the Fiedler value and Fiedler vector related to the graph in Figure 1.26 and cluster the graph into two subgraphs using these values.

9. A subpopulation of marmots is displaying a dominance behavior represented by the directed graph shown in Figure 1.27. The weight, w_{ij} on the edge from marmot i to marmot j represents the number of times that marmot i showed a dominance behavior over marmot j.

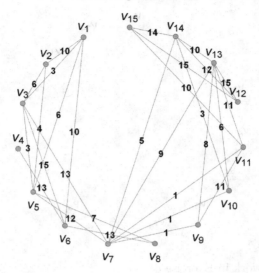

FIGURE 1.26
Weighted marmot social network in Exercise 8.

  a. Use a score vector

$$\vec{S} = \frac{s_{ij}}{n_i}$$

similar to that used in the sports ranking in Section 1.6, where n_i

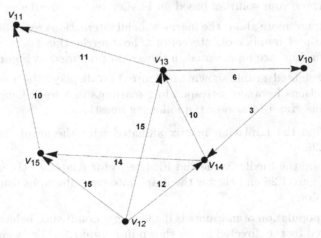

FIGURE 1.27
Weighted dominance interactions in Exercise 9.

represents the number of interactions with marmot i and s_{ij} represents the number of dominant interactions between marmots i and j noting that if marmot j is dominant over marmot i then s_{ij} is negative.

b. If A is the adjacency matrix representing the graph in Figure 1.27, use $A + I$ and S to determine a ranking based on 1st and 2nd generation dominance.

c. Use $A + I$ and S, to determine a ranking based on 1st through 3rd generation dominance.

10. The Cinderella data from GitHub link 1 is a binary data set of traits in which 31 Cinderella tales from around the world contain. Use this data to answer the following questions.

a. Find the Gramian matrix, S.

b. What do the (1,1) and (1,2) entries represent in S?

c. Find the Fiedler vector related to the Gramian matrix, S, and use this information to cluster the Cinderella tales into two clusters.

11. Sports Ranking in Section 1.6 uses a weighted system where 2 or $3 - step$ wins between teams carry equal weight to direct head to head competition wins in order to create a ranking from

$$\sum_{i=0}^{n} A_{SEC*}^{i} \vec{S}.$$

If instead, $N - step$ wins carried a weight of $\frac{1}{N}$ and thus the ranking would be

$$\vec{S} + \sum_{i=1}^{n} \frac{1}{i} A_{SEC*}^{i} \vec{S}.$$

determine the ranking of the teams in the tournament from Section with $n = 4$ and $n = 5$.

2

Stochastic Processes

The goal of this chapter is to use matrices to explore the behavior of systems as they change, or transition, throughout time. In order to do so, we begin by defining some properties of matrices that will allow us to employ some important theories, such as the *Perron-Frobenius Theorem*.

Reducibility

A **permutation** is a change of ordering. So for example, if the numbers (1, 2, 3, 4) are rearranged, or permuted, to be read as (2, 1, 4, 3) we see that

$$1 \rightarrow 2, 2 \rightarrow 1, 3 \rightarrow 4, \text{ and } 4 \rightarrow 3.$$

We denote this permutation with the notation (12)(34). Similarly the permutation denoted by (1)(234) represents

$$1 \rightarrow 1, 2 \rightarrow 3, 3 \rightarrow 4, \text{ and } 4 \rightarrow 2$$

and the ordering of the four numbers is (1, 3, 4, 2).

A **permutation matrix**, P, is a matrix in which the rows of the identity matrix of the same dimensions are permuted.

Example 2.1. *A permutation matrix can permute one or more rows of the identity matrix. Let*

$$P_1 = \begin{pmatrix} 1 & 0 & 0 & 0 \\ 0 & 0 & 0 & 1 \\ 0 & 0 & 1 & 0 \\ 0 & 1 & 0 & 0 \end{pmatrix} \text{ and } P_2 = \begin{pmatrix} 1 & 0 & 0 & 0 \\ 0 & 0 & 1 & 0 \\ 0 & 0 & 0 & 1 \\ 0 & 1 & 0 & 0 \end{pmatrix}.$$

DOI: 10.1201/9781003025672-2

The row permutations related to P_1 and P_2 respectively are $(1)(3)(42)$ and $(1)(234)$. Given another matrix

$$
A = \begin{pmatrix} 1 & 2 & 3 & 4 \\ -1 & 0 & 1 & 2 \\ -2 & -1 & 0 & 1 \\ -3 & -2 & -1 & 0 \end{pmatrix}
$$

$$
P_1 A = \begin{pmatrix} 1 & 2 & 3 & 4 \\ -3 & -2 & -1 & 0 \\ -1 & 0 & 1 & 2 \\ -2 & -1 & 0 & 1 \\ -1 & 0 & 1 & 2 \end{pmatrix}
$$

$$
P_2 A = \begin{pmatrix} 1 & 2 & 3 & 4 \\ -3 & -2 & -1 & 0 \\ -1 & 0 & 1 & 2 \\ -2 & -1 & 0 & 1 \end{pmatrix}
$$

we can see how the respective permutation matrices transform the matrix A. Recall that permuting two rows in an identity matrix is an elementary row operation, and the affiliated elementary matrix has a determinant of -1. With this in mind,

$$
|P_1| = -1, \ |P_2| = 1, \ and \ |P_2 A| = |P_2||A| = |A|.
$$

A $n \times n$ matrix A is called **reducible** if there exists a permutation matrix P such that

$$
PAP^T = \begin{pmatrix} A_{11} & A_{12} \\ \hline 0 & A_{22} \end{pmatrix}
$$

where A_{11} is a $r \times r$ matrix and A_{12} and A_{22} are $(n-r) \times (n-r)$ matrices. If a matrix is not reducible then it is called **irreducible**.

Notice that a matrix A is reducible if and only if a permutation matrix can be found to put A into a block upper triangular form.

Example 2.2.

$$A_1 = \left(\begin{array}{cc|cc} 0 & 1 & 0 & 0 \\ 0 & 0 & 1 & 0 \\ \hline 0 & 0 & 0 & 1 \\ 0 & 0 & 0 & 0 \end{array} \right)$$

$$A_2 = \left(\begin{array}{cccc} 0 & 1 & 0 & 1 \\ 0 & 0 & 0 & 0 \\ 0 & 1 & 0 & 0 \\ 0 & 0 & 1 & 0 \end{array} \right)$$

$$A_3 = \left(\begin{array}{cccc} 0 & 1 & 0 & 1 \\ 0 & 0 & 0 & 0 \\ 0 & 1 & 0 & 1 \\ 0 & 0 & 0 & 0 \end{array} \right)$$

Matrix A_1 is reducible as it is already in the form of a block upper triangular matrix. In its original form, matrix A_2 is not block upper triangular; however, there is a permutation matrix

$$P = \left(\begin{array}{cccc} 1 & 0 & 0 & 0 \\ 0 & 0 & 0 & 1 \\ 0 & 0 & 1 & 0 \\ 0 & 1 & 0 & 0 \end{array} \right) \quad \text{such that}$$

$$PA_2P^T = \left(\begin{array}{cc|cc} 0 & 1 & 0 & 1 \\ 0 & 0 & 1 & 0 \\ \hline 0 & 0 & 0 & 1 \\ 0 & 0 & 0 & 0 \end{array} \right)$$

which is block upper triangular. Unlike the other two matrices, matrix A_3 is irreducible.

Theorem 5. *A symmetric matrix is irreducible if and only if its associated graph is connected. A matrix is irreducible if and only if its associated graph is* **strongly-connected**, *that is, any pair of vertices has a path between them.*

Notice that the graphs affiliated with the adjacency matrices A_1 and A_2 in Example 2.2, shown in Figure 2.1, are strongly connected while the graph affiliated with matrix A_3 is not strongly connected since there is not a path between vertices v_1 and v_3 or v_2 and v_4.

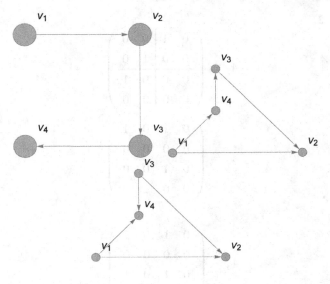

FIGURE 2.1
Graphs affiliated with Example 2.2.

A square matrix, A, is said to be a **non-negative matrix**, denoted $A \geq 0$, if all of its entries are non-negative, and a **positive matrix**, denoted $A > 0$, if all of its entries are positive.

A matrix, A, is said to be a **primitive matrix** if there exist integer $k > 0$ such that A^k is a positive matrix. A matrix is irreducible if for all (i,j) there exist a k such that $A_{i,j}^k > 0$.

Lemma 2. *If a square non-negative matrix A is primitive then it is irreducible.*

Example 2.3. *Let*

$$A = \begin{pmatrix} 0 & 0 & 1 \\ 3 & 4 & 1 \\ 3 & 2 & 4 \end{pmatrix}.$$

Then,

$$A^2 = \begin{pmatrix} 3 & 2 & 4 \\ 15 & 17 & 29 \\ 14 & 14 & 27 \end{pmatrix}.$$

Notice that $A_{ij}^2 > 0$ for $1 \leq i,j \leq n$ so A is primitive and irreducible.

The following theorem, the Perron-Frobenius theorem, is fundamental in the study of discrete probabilistic behaviors, also called **stochastic processes**, in relation to eigenvalues and eigenvectors of irreducible matrices.

Theorem 6. *(Perron-Frobenius Thereom) If A is an $n \times n$ non-negative irreducible matrix then*

- *One of its eigenvalues, λ_{PF} with multiplicity 1, is greater than or equal to, in absolute value, all of the other eigenvalues.*

- *There is a positive eigenvector corresponding to λ_{PF}.*

Perron is also credited with a slightly different theorem that relies on a matrix being primitive rather than irreducible.

Theorem 7. *(Perron Theorem)[4]*
If A is a primitive matrix with the largest eigenvalue in magnitude, the spectral radius, $\rho(A) = \lambda_{PF}$, then λ_{PF} is an eigenvalue of multiplicity 1 and has a corresponding strictly positive eigenvector.

In the case when a matrix is non-negative but not primitive or irreducible, then the multiplicity of λ_{PF} may be greater than 1.

Example 2.4. *Let $A_1 = \begin{pmatrix} 1 & 1 & 1 \\ 1 & 0 & 0 \\ 0 & 1 & 1 \end{pmatrix}$. Notice that*

$$A_1^2 = \begin{pmatrix} 2 & 2 & 2 \\ 1 & 1 & 1 \\ 1 & 1 & 1 \end{pmatrix}$$

and thus A_1 is primitive. From the Perron Theorem, we can conclude that

$$\lambda_{PF} = 2 \text{ is an eigenvalue with multiplicity 1}$$

and all other eigenvalues are smaller in magnitude then λ_{PF}.

Applying the Perron-Frobenius Theorem to the matrix A in Example 2.3, $\lambda_{PF} = \rho(A) = 6$ with corresponding unit eigenvector

$$\left\{ \frac{2}{\sqrt{229}}, \frac{9}{\sqrt{229}}, \frac{12}{\sqrt{229}} \right\}.$$

Similarly, the crime network matrix presented in Figure 1.8 is also irreducible with $\lambda_{PF} = 4.032$ and with a corresponding unit eigenvector

$$\{0.267, 0.305, 0.163, 0.285, 0.353, 0.324, 0.272, 0.348, 0.322, 0.447\}.$$

2.1   Markov Chain Basics

A **stochastic process** is a sequence of random variables indexed over time, either continuous or discrete. We consider here a discrete stochastic process referred to as a **Markov chain** , in which

$$P(X_n = x_n | X_1 = x_1, \ \ldots, \ X_{n-1} = x_{n-1}) = P(X_n = x_n | X_{n-1} = x_{n-1}).$$

That is, that the probability of the current state is only dependent on the previous states; there have been no other states that have occurred over time. This property is referred to as the **memoryless property**.

A Markov chain has a corresponding **transition matrix**. The transition matrix, M, is a probability matrix, where $M_{i,j}$ is the probability of going from state j to state i,

$$M_{i,j} = P(X_n = i, \ X_{n-1} = j).$$

Example 2.5. *An example of a transition matrix, M, for a Markov Chain,*

$$
\begin{array}{cc}
 & \text{Preceding State} \\
New\ State \begin{array}{c} 1 \\ 2 \\ 3 \end{array} & \left(\begin{array}{ccc} 1 & 2 & 3 \\ 0.05 & 0.7 & 0.46 \\ 0.75 & 0.2 & 0.12 \\ 0.2 & 0.1 & 0.42 \end{array} \right)
\end{array}.
$$

It is sometimes helpful in a Markov chain to visualize the system with a state diagram, with arrows between states representing the transitions and weights representing the probability of that transition.

An example of a state diagram corresponding to the transition matrix can be seen in Figure 2.2.

There are a few ways that we can initially start to analyze the long-term behavior of the states given a transition matrix.

One way is to make an assumption about the initial value of each state, creating an initial vector, π. Then the long-term behavior can be found by applying powers of the transition matrix to π,

$$lim_{k \to \infty} M^k \pi.$$

Theorem 8. *Given a Markov chain $X_1, X_2, \ldots, X_k$ with transition matrix M,*

$$v = lim_{n \to \infty} X_n$$

where v is the unit eigenvector associated with eigenvalue $\lambda_{PF} = 1$.

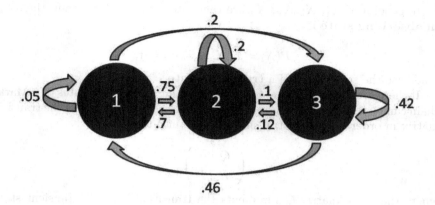

FIGURE 2.2
Example of a state diagram.

Proof. The Perron-Frobenius theorem is the main tool in this proof. If the Markov chain has one or more absorbing states, the transition matrix can be written in a form such that the absorbing states are all in the last rows (and columns). In this case, the transition matrix is irreducible. However, if all states in the Markov chain are transient, the transition matrix may be irreducible or primitive, or may only be non-negative. Thus, we will only make the assumption that the transition matrix, M, is non-negative. $\qquad\square$

Using the general case of the Perron-Frobenius Theorem, we assume that the largest eigenvalue in magnitude λ_{PF} has a corresponding unit eigenvector v.

If M is an $n \times n$ transition matrix with entries $M_{i,j}$, then

$$\lambda_{PF}v = Mv = \sum_{j=1}^{n} M_{i,j}v_j.$$

Thus $\lambda_{PF}vv^T = \sum_{j=1}^{n} M_{i,j} = 1$ and $\lambda_{PF} = 1$. In addition,

$$lim_{n\to\infty} X_n = lim_{n\to\infty} M^n v = lim_{n\to\infty} \lambda_{PF}^n v = v.$$

The vector $\frac{v}{\sum_i v_i}$ will provide a percent vector for the long-run behavior of the Markov chain.

Example 2.6. *For the transition matrix presented in Example 2.5, the percent eigenvector corresponding with $\lambda_{PF} = 1$ is $\{.39, .40, .21\}$. From this eigenvector, we can determine the long-run percentages for each state.*

In general, if $X_1, X_2, \ldots, X_n$ are the states in a Markov chain, then X_i is an **absorbing state** if

$$P(X_i = x_i | X_{i-1} = x_i) = 1,$$

otherwise the state is called a **transient state**.

If there are q transient states and r absorbing states, the states of a Markov chain can be renumbered, thus permuting the rows and columns of the transition matrix in order to put the transition matrix into the form

$$\begin{pmatrix} Q & 0 \\ R & I \end{pmatrix},$$

where the $q \times q$ matrix Q represents the transition between transient states, the $r \times r$ matrix I represents absorption, the $q \times r$ matrix R represents the transition from a transient state to an absorbing state, and the $r \times q$ matrix 0 represents the lack of ability to move from an absorbing state to a transient state.

By putting the transition matrix into this form, we can employ another important matrix in the study of Markov chains, the **fundamental matrix**,

$$N = \sum_{n=0}^{\infty} Q^n.$$

Recall from your study of series that the Taylor series with center 0, also called the Maclaurin series, for the function $\frac{1}{1-x}$ is $\sum_{n=0}^{\infty} x^n$. Similarly, if a Markov chain has at least one absorbing state, then $I - Q$ is invertible with inverse $N = \sum_{n=0}^{\infty} Q^n$.

The $(i,j)^{\text{th}}$ entry of the fundamental matrix, N, represents the expected time that the Markov chain stays in state X_j given that it starts in X_i.

Example 2.7. *Assume that you are in an escape room maze with 5 rooms. You are able to escape from either room 4 or 5 and your ability to transition from one room to the next is given by the state diagram in Figure 2.3.*

You wish to determine how long you should expect it to take if you escape through room 5 given that you start in room 1.

In this example, the transition matrix,

$$M = \left(\begin{array}{ccc|cc} \frac{1}{3} & \frac{2}{5} & 0 & 0 & 0 \\ \frac{2}{3} & \frac{1}{3} & \frac{3}{8} & 0 & 0 \\ 0 & \frac{1}{5} & \frac{1}{2} & 0 & 0 \\ 0 & \frac{1}{15} & 0 & 1 & 0 \\ 0 & 0 & \frac{1}{8} & 0 & 1 \end{array} \right),$$

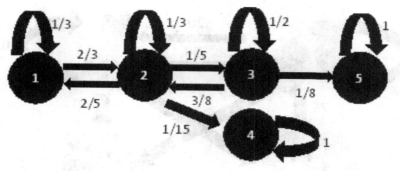

FIGURE 2.3
State diagram representing the transition between rooms in the maze.

$$Q = \begin{pmatrix} \frac{1}{3} & \frac{2}{5} & 0 \\ \frac{2}{3} & \frac{1}{3} & \frac{3}{8} \\ 0 & \frac{1}{5} & \frac{1}{2} \end{pmatrix},$$

$$and \ N = \begin{pmatrix} \frac{93}{14} & \frac{36}{7} & \frac{27}{7} \\ \frac{60}{7} & \frac{60}{7} & \frac{45}{7} \\ \frac{24}{7} & \frac{24}{7} & \frac{32}{7} \end{pmatrix}.$$

From matrix N, we can conclude that you should be expected to take

$$\frac{93}{14} + \frac{60}{7} + \frac{24}{7} = \frac{261}{14}.$$

So on average, roughly 19 time steps are needed if you wish to start in room 1 and escape from room 5.

2.2 Hidden Markov Models

Markov chains are extremely helpful when we wish to determine a sequence of events based on probabilities of observed behaviors. However, there are times when we wish to make predictions but these probabilities are not already clear or observed. In this case, we say that we have a **Hidden Markov Model, HMM**. We begin with an example,

Example 2.8. *Kalu and Rath own a small start-up day trading company. Rath tells Kalu that if the stock market is bearish, on the decline, on any given day then she will be unhappy; however, if the market is bullish, on the rise, then she will be happy.*

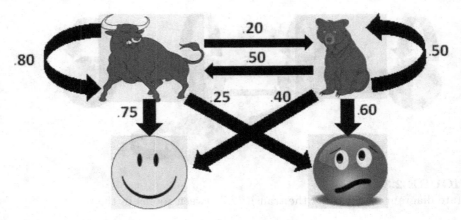

FIGURE 2.4
Stock Market State Diagram Depicting a HMM.

If Rath calls Kalu and tells him that last week she was

{Happy, Sad, Happy, Happy, Sad},

then Kalu would assume based on this that the stock market was

{Bullish, Bearish, Bullish, Bullish, Bearish}.

Unfortunately, the health of the stock market alone may not be the only factor weighing on Rath's daily mood. Perhaps Rath is actually happy 75% of the time when the market is bullish and sad otherwise, and when the stock market is bearish, she is sad 60% of the time and otherwise happy.

Additionally, let's assume that we know additional information about daily changes in the stock market that are shown in the state diagram in Figure 2.4.

Under these new conditions, if Rath tells Kalu that last week she was {Happy, Sad, Happy, Happy, Sad}, Kalu might want to determine how likely his assumption is that the market status was {Bullish, Bearish, Bullish, Bullish, Bearish}.

In this example, there are observations, Rath's daily moods, and hidden states, which are the daily stock market trends.

*Like traditional Markov chains, this model has transition probabilities, which are those probabilities between the hidden states, the stock market states in this example. Additionally, a hidden Markov model has **emissions probabilities**, which are those probabilities emitted from the hidden states to the observations.*

The question posed in Example 2.8 is one of three common questions that are addressed by hidden Markov models. These questions fit into three general categories

1. **The Likelihood Problem**—determining the probability of an observation sequence given a model.

2. **The Decoding Problem**—determining the most likely sequence of events given an observation sequence.

3. **The Learning Problem**—determining the most likely model (with parameters) given the observation sequence.

2.2.1   The Likelihood Problem

Because we will have to use prior information to make some inferences in a hidden markov model, a strong understanding of **Bayes Theorem** is essential.

Two events, X_1 and X_2, are **mutually exclusive** if

$$X_1 \cap X_2 = \emptyset.$$

Events $X_1, X_2, \ldots, X_n$ are **exhaustive** if

$$\cup_{i=1}^n X_i = S,$$

where S is the set of all possible outcomes.

Example 2.9. *Assume that within a population in which COVID-19 variants are prevalent, a random human may be*

$$X_1 = susceptible, \; X_2 = infected,$$

or $X_3 = recovered$ and temporarily immune.

If contraction of one variant provides temporary immunity for all variants, these sets can be considered mutually exclusive since

$$X_1 \cap X_2 = X_1 \cap X_3 = X_2 \cap X_3 = \emptyset.$$

Additionally X_1, X_2, and X_3 are exhaustive events. However, if contractions of one variant only provides temporary immunity to that variant, then $X_1 \cap X_3 \neq \emptyset$ and X_1 and X_3 are not mutually exclusive events.

Theorem 9. *(Baye's Theorem) Let $X_1, X_2, \ldots, X_n$ be mutually exclusive and exhaustive events*

$$P(X_j | X_k) = \frac{P(X_k | X_j) P(X_j)}{\sum_{i=1}^n P(X_k | X_i) P(X_i)}.$$

Example 2.10. *Given the state diagram from Example 2.8, Kalu looked online to determine that in recent history the stock market was bullish 87% of the time and bearish the other 13% of the time. He used this information to determine the probability that the stock market is bullish on a given day if he knows Rath is happy.*

$$P(bullish|happy) = \frac{P(happy|bullish)P(bullish)}{P(happy|bullish)P(bullish) + P(happy|bearish)P(bearish)}$$

$$= \frac{.75 \cdot .87}{.75 \cdot .87 + .4 \cdot .13} \approx .93$$

Similarly, Kalu can determine the probability that the stock market is bearish on a given day if he knows Rath is happy.

$$P(bearish|happy) = \frac{P(happy|bearish)P(bearish)}{P(happy|bullish)P(bullish) + P(happy|bearish)P(bearish)}$$

$$= \frac{.4 \cdot .13}{.75 \cdot .87 + .4 \cdot .13} \approx .07$$

Finishing off his analysis, Kalu wants to determine

$$P(bullish|sad) = \frac{P(sad|bullish)P(bullish)}{P(sad|bullish)P(bullish) + P(sad|bearish)P(bearish)}$$

$$= \frac{.25 \cdot .87}{.25 \cdot .87 + .6 \cdot .13} \approx .74$$

$$P(bearish|sad) = \frac{P(sad|bearish)P(bearish)}{P(sad|bullish)P(bullish) + P(sad|bearish)P(bearish)}$$

$$= \frac{.6 \cdot .13}{.25 \cdot .87 + .6 \cdot .13} \approx .26$$

So Kalu is much more confident in saying that the stock market is Bullish given that Rath is Happy than he is saying that the stock market is Bearish given that Rath is Sad.

Given that Rath tells Kalu that she is {Happy, Sad, Happy, Happy, Sad} the likelihood of the market being {Bullish, Bearish, Bullish, Bullish, Bearish} is

$$.93 \cdot .25 \cdot .93 \cdot .93 \cdot .26 \approx .05.$$

Like we saw in Example 2.10, finding the probability of an observation sequence given a model is a type of likelihood problem. The **Forward/Backward Algorithm** is an algorithm to solve this in general using hidden Markov models.

Algorithm 1. *Forward Algorithm*

Let O be an observed sequence of random variables, $O_1, O_2, \ldots, O_n$ and H be the hidden sequence of random variables, $H_1, H_2, \ldots, H_n$ and λ is the hidden Markov model with initial probabilities, π. We begin by finding the joint probability distribution

$$P(O_1, O_2, \ldots, O_n|\lambda) = \sum_k P(O_1, O_2, \ldots, O_n, H_n = k|\lambda) = \sum_k \alpha(H_n)^k,$$

where

$$\alpha(H_t)^k = P(O_1, O_2, \ldots, O_t, H_t = k|\lambda), \quad \text{for all } k$$
$$= P(O_t|H_t = k, \lambda) \sum_i P(H_t = k|H_{t-1} = i)\alpha(H_{t-1})^i,$$

Note that

$$\alpha(H_1)^k = P(O_1|H_1 = k)P(H_1 = k)$$

for all k, and that H_0 is the start of the sequence.

The Forward Algorithm can be used to determine the probability of an observed sequence given the occurrence of a hidden sequence, $P(O|H)$, and the likelihood of an observed sequence, $P(O)$.

$$\alpha(H_t) = E_{(O_t, \cdot)} \circ (T\alpha(H_{t-1}))$$
$$P(O_1, O_2, \ldots, O_t) = E_{(O_t, \cdot)}(T\alpha(H_{t-1}))$$

for all t where $\alpha(H_0) = \pi$.

Example 2.11. *(Forward Algorithm) Continuing Example 2.10, Kalu can use the Forward Algorithm to determine the likelihood of his observation of Rath's mood*

$$P(O) = P(O_1 = Happy, O_2 = Sad, O_3 = Happy, O_4 = Happy, O_5 = Sad).$$

Kalu can put all of the information he has found into two matrices thus far, the transition matrix, T, and the emission matrix, E.

The entries of the transition matrix T are

$$T_{i,j} = P(H_t = state \ i|H_{t-1} = state \ j).$$

For this example,

$$T = \begin{matrix} \text{\tiny Next state} \end{matrix} \begin{matrix} Bullish \\ Bearish \end{matrix} \begin{pmatrix} \overset{\textstyle Preceding\ state}{\overset{Bullish \quad Bearish}{\begin{matrix} .8 & .5 \\ .2 & .5 \end{matrix}}} \end{pmatrix}.$$

The emissions matrix, E, is a transition matrix between hidden and observed states, where

$$E_{i,j} = P(O_t = state\ i | H_t = state\ j).$$

Kalu has determined that the emission matrix is

$$E = \begin{matrix} \text{\tiny Observed state} \end{matrix} \begin{matrix} Happy \\ Sad \end{matrix} \begin{pmatrix} \overset{\textstyle Hidden\ state}{\overset{Bullish \quad Bearish}{\begin{matrix} .75 & .40 \\ .25 & .60 \end{matrix}}} \end{pmatrix}.$$

If $\pi = \begin{pmatrix} P(Bullish) \\ P(Bearish) \end{pmatrix} = \begin{pmatrix} .87 \\ .13 \end{pmatrix},$

$$\alpha(H_1)^{Bull} = P(O_1 = Happy | H_1 = Bull)P(H_1 = Bull) = .87 \cdot .75,$$

$$\alpha(H_1)^{Bear} = P(O_1 = Happy | H_1 = Bear)P(H_1 = Bear) = .13 \cdot .40,$$

$$\alpha(H_1) = \begin{pmatrix} \alpha(H_1)^{Bull} \\ \alpha(H_1)^{Bear} \end{pmatrix} = \begin{pmatrix} .6525 \\ .052 \end{pmatrix}.$$

$$P(O_1 = Happy) = \sum_k \alpha(H_1)^k = E_{(O_1 = Happy, \cdot)}\pi = .6525 + .052 = .7045.$$

$$\alpha(H_2)^{Bull} = P(O_2 = Sad | H_2 = Bull)(P(H_2 = Bull | H_1 = Bull)\alpha(H_1)^{Bull}$$
$$+ P(H_2 = Bull | H_1 = Bear)\alpha(H_1)^{Bear})$$
$$= .25 \cdot T_{(Bull, \cdot)}\alpha(H_1) = .25 \cdot (.8 \cdot .6525 + .5 \cdot .052)$$

$$\alpha(H_2)^{Bear} = P(O_2 = Sad | H_2 = Bear)(P(H_2 = Bear | H_1 = Bull)\alpha(H_1)^{Bull}$$
$$+ P(H_2 = Bear | H_1 = Bear)\alpha(H_1)^{Bear})$$
$$= .6 \cdot T_{(Bear, \cdot)}\alpha(H_1) = .60 \cdot (.2 \cdot .6525 + .5 \cdot .052)$$

Another way to determine,

$$\alpha(H_2) = E_{(O_2 = Sad, \cdot)} \circ (T\alpha(H_1))$$

$$= \begin{pmatrix} .25 \\ .60 \end{pmatrix} \circ \left(\begin{pmatrix} .8 & .5 \\ .2 & .5 \end{pmatrix} \begin{pmatrix} .6525 \\ .052 \end{pmatrix} \right) = \begin{pmatrix} .137 \\ .0939 \end{pmatrix}$$

$$P(O_1 = Happy, O_2 = Sad) = \sum_k \alpha(H_2)^k = E_{(Sad, \cdot)}(T\alpha(H_1)) = .2309$$

$$\begin{aligned} \alpha(H_3)^{Bull} &= P(O_3 = Happy | H_3 = Bull)(P(H_3 = Bull | H_2 = Bull)\alpha(H_2)^{Bull} \\ &+ P(H_3 = Bull | H_2 = Bear)\alpha(H_2)^{Bear}) \\ &= .75 \cdot (.8 \cdot .137 + .5 \cdot .094) \end{aligned}$$

$$\begin{aligned} \alpha(H_3)^{Bear} &= P(O_3 = Happy | H_3 = Bear)(P(H_3 = Bear | H_2 = Bull)\alpha(H_2)^{Bull} \\ &+ P(H_3 = Bear | H_2 = Bear)\alpha(H_2)^{Bear}) \\ &= .40 \cdot (.2 \cdot .137 + .5 \cdot .094) \end{aligned}$$

$$\alpha(H_3) = E_{(Happy, \cdot)} \circ (T\alpha(H_2)) = \begin{pmatrix} .11741 \\ .02974 \end{pmatrix}$$

$$P(O_1 = Happy, O_2 = Sad, O_3 = Happy) = \sum_k \alpha(H_3)^k = E_{(Happy, \cdot)}(T\alpha(H_2))$$

$$= .14715$$

$$\alpha(H_4) = \begin{pmatrix} 0.0816 \\ 0.01534 \end{pmatrix}, \; P(O_1 = Happy, \emptyset_2 = Sad, \emptyset_3 = Happy, O_4 = Happy)$$

$$= .09693$$

$$\alpha(H_5) = \begin{pmatrix} 0.0182 \\ 0.0144 \end{pmatrix}. \; Thus, \; P(O | \lambda) = .0326$$

Now that we have used the Forward Algorithm to compute the probability of the joint observed states, we can use the **Backward Algorithm** to determine marginal probabilities $P(H_t = k | O)$.

Algorithm 2. *Forward/Backward Algorithm*

Let O be an observed sequence of random variables, $O_1, O_2, \ldots, O_n$ and H be the hidden sequence of random variables, $H_1, H_2, \ldots, H_n$.

Compute $\alpha(H_t)^k$ for $t = 1, \ldots, n$ as defined in the Forward Algorithm. Then work backward, where

$$\beta(H_n)^k = 1$$

for all k and define

$$\beta(H_t)^k = P(O_{t+1}, O_{t+2}, \ldots, O_n | H_t = k, \lambda) \quad \text{for all } k$$
$$= \sum_i P(O_{t+1}|H_{t+1} = i)P(H_{t+1} = i|H_t = k)\beta(H_{t+1})^i.$$

Using matrix operations,

$$\beta(H_{t-1}) = T^T(E_{(O_t, \cdot)} \circ \beta(H_t))$$

where $E_{(O_t, \cdot)}$ is the row affiliated with O_t in the emissions matrix E.

We now can think about determining $P(O|\lambda)$ using a variety of different tools.

$$P(O|\lambda) = \sum_k \alpha(H_n)^k = \sum_k E_{(O_1, k)}\beta(H_1)^k \pi_k.$$

The Forward/Backward Algorithm can be applied in many of the different hidden markov model questions.

We begin by seeing how the Backward Algorithm can be applied in much the same way as the Forward Algorithm to determine $P(O|\lambda)$.

Example 2.12. *(Backward Algorithm)*
Recall the transition and emission matrices, T and E, from Example 2.11,

$$T = \begin{array}{c} \text{Next state} \end{array} \begin{array}{c} \\ Bull \\ Bear \end{array} \begin{pmatrix} \begin{array}{c} Preceding\ state \\ Bull \quad Bear \end{array} \\ .8 \qquad .5 \\ .2 \qquad .5 \end{pmatrix}, \quad E = \begin{array}{c} \text{Observed state} \end{array} \begin{array}{c} \\ Happy \\ Sad \end{array} \begin{pmatrix} \begin{array}{c} Hidden\ state \\ Bull \quad Bear \end{array} \\ .75 \qquad .40 \\ .25 \qquad .60 \end{pmatrix}.$$

We begin by setting $\beta(H_5)^{Bull} = \beta(H_5)^{Bear} = 1$.

$$\beta(H_4)^{Bull} = P(O_5 = Sad|H_5 = Bull)P(H_5 = Bull|H_4 = Bull)\beta(H_5)^{Bull}$$
$$+ P(O_5 = Sad|H_5 = Bear)P(H_5 = Bear|H_4 = Bull)\beta(H_5)^{Bear}$$
$$= .25 \cdot .8 \cdot 1 + .6 \cdot .2 \cdot 1 = .32$$

$$\beta(H_4)^{Bear} = P(O_5 = Sad|H_5 = Bull)P(H_5 = Bull|H_4 = Bear)\beta(H_5)^{Bull}$$
$$+ P(O_5 = Sad|H_5 = Bear)P(H_5 = Bear|H_4 = Bear)\beta(H_5)^{Bear}$$
$$= .25 \cdot .5 \cdot 1 + .6 \cdot .5 \cdot 1 = .425$$

Another way to think about this is that,

$$\beta(H_4) = T^T(E_{(O_5=Sad,\cdot)} \circ \beta(H_5))$$

$$= \begin{pmatrix} .8 & .2 \\ .5 & .5 \end{pmatrix} \left(\begin{pmatrix} .25 \\ .60 \end{pmatrix} \circ \begin{pmatrix} 1 \\ 1 \end{pmatrix} \right) = \begin{pmatrix} .32 \\ .425 \end{pmatrix}$$

$$\beta(H_3)^{Bull} = P(O_4 = Happy|H_4 = Bull)P(H_4 = Bull|H_3 = Bull)\beta(H_4)^{Bull}$$
$$+ P(O_4 = Happy|H_4 = Bear)P(H_4 = Bear|H_3 = Bull)\beta(H_4)^{Bear}$$
$$= .75 \cdot .8 \cdot .32 + .4 \cdot .2 \cdot .425 = .226$$

$$\beta(H_3)^{Bear} = P(O_4 = Happy|H_4 = Bull)P(H_4 = Bull|H_3 = Bear)\beta(H_4)^{Bull}$$
$$+ P(O_4 = Happy|H_4 = Bear)P(H_4 = Bear|H_3 = Bear)\beta(H_4)^{Bear}$$
$$= .75 \cdot .5 \cdot .325 + .4 \cdot .5 \cdot .42 = .205$$

$$\beta(H_3) = T^T(E_{(O_4=Happy,\cdot)} \circ \beta(H_4)) = \begin{pmatrix} .8 & .2 \\ .5 & .5 \end{pmatrix} \left(\begin{pmatrix} .75 \\ .40 \end{pmatrix} \circ \begin{pmatrix} .32 \\ .425 \end{pmatrix} \right)$$

$$= \begin{pmatrix} .226 \\ .205 \end{pmatrix}$$

$$\beta(H_2) = T^T(E_{(O_3=Happy,\cdot)} \circ \beta(H_3)) = \begin{pmatrix} .8 & .2 \\ .5 & .5 \end{pmatrix} \left(\begin{pmatrix} .75 \\ .40 \end{pmatrix} \circ \begin{pmatrix} .226 \\ .205 \end{pmatrix} \right)$$

$$= \begin{pmatrix} .152 \\ .12575 \end{pmatrix}$$

$$\beta(H_1) = T^T(E_{(O_2=Sad,\cdot)} \circ \beta(H_2)) = \begin{pmatrix} .8 & .2 \\ .5 & .5 \end{pmatrix} \left(\begin{pmatrix} .25 \\ .60 \end{pmatrix} \circ \begin{pmatrix} .152 \\ .12575 \end{pmatrix} \right)$$

$$= \begin{pmatrix} 0.04549 \\ 0.056725 \end{pmatrix}$$

If $\pi_k = (.87, .13)$,

$$P(O|\lambda) = \sum_k E_{(Happy,k)}\beta(H_1)^k \pi_k$$

$$= .75 \cdot .04549 \cdot .87 + .4 \cdot .056725 \cdot .13 = .0326$$

2.2.2   The Decoding Problem

In solving the decoding problem with a HMM, we wish to find the most likely hidden sequence H^* of occurring given an observation sequence, O. More specifically, the goal of decoding is to determine

$$H^* = argmax_H P(H|O).$$

Note that $argmax_H P(H|O)$ is the maximum probability of a hidden sequence, H, over all possible $H's$, given an observation sequence O.

We will employ both the Forward/Backward Algorithm and the **Viterbi Algorithm** in our investigation of the decoding problem.

Applying the Forward/Backward Algorithm

Define

$$\gamma(H_t)^k = P(H_t = k|O,\lambda)$$
$$= \frac{P(H_t = k, O|\lambda)}{P(O|\lambda)}$$
$$= \frac{\alpha(H_t)^k \beta(H_t)^k}{P(O|\lambda)}$$
$$= \frac{\alpha(H_t)^k \beta(H_t)^k}{\alpha(H_n) \cdot \beta(H_n)}$$

Example 2.13. *Recall from Example 2.11 that* $\alpha(H_1) = \begin{pmatrix} .6525 \\ .052 \end{pmatrix}$ *and from Example 2.12 that* $\beta(H_1) = \begin{pmatrix} .04549 \\ 0.056725 \end{pmatrix}$.

Kalu can use this information to determine the likelihood of the hidden state sequence $H = H_1, H_2, \ldots, H_n$.

$$P(H_1 = Bullish|O,\lambda) = \gamma(H_1)^{Bull}$$
$$= \frac{\alpha(H_1)^{Bull} \beta(H_1)^{Bull}}{\alpha(H_1) \cdot \beta(H_1)} = .91$$

and

$$P(H_1 = Bearish|O,\lambda) = \gamma(H_1)^{Bear}$$

$$= \frac{\alpha(H_1)^{Bear}\beta(H_1)^{Bear}}{\alpha(H_1)\cdot\beta(H_1)} = .09$$

In a similar manner, Kalu determines that

$$P(H_2 = Bullish|O,\lambda) = \gamma(H_2)^{Bull} = .638$$
$$P(H_2 = Bearish|O,\lambda) = \gamma(H_2)^{Bear} = .362$$
$$P(H_3 = Bullish|O,\lambda) = \gamma(H_3)^{Bull} = .81$$
$$P(H_3 = Bearish|O,\lambda) = \gamma(H_3)^{Bear} = .19$$
$$P(H_4 = Bullish|O,\lambda) = \gamma(H_4)^{Bull} = .80$$
$$P(H_4 = Bearish|O,\lambda) = \gamma(H_4)^{Bear} = .20$$
$$P(H_5 = Bullish|O,\lambda) = \gamma(H_5)^{Bull} = .56$$
$$P(H_5 = Bearish|O,\lambda) = \gamma(H_5)^{Bear} = .44$$

Similar to the Forward/Backward Algorithm, the Viterbi Algorithm still requires knowledge of the the transition matrix, T, and the emissions matrix, E, as well as π, the initial probability distribution of the hidden random variables. We begin by thinking through the recursive idea of Viterbi Algorithm.

Assume that the observed sequence of random variables is $O_1, O_2, \ldots, O_n$ and the hidden sequence of random variables is $H_1, H_2, \ldots, H_n$.

An additional important fact is that the conditional probability $P(H|O)$ is proportional to the joint probability $P(H,O)$,

$$P(H|O)P(O) = P(H,O)$$

and thus if we find the hidden sequence that maximizes the joint probability, $P(H,O)$, this sequence will also maximize the conditional probability, $P(H|O)$.

$$H^* = argmax_{H_i,1\leq i\leq n}P(H_i|O_i)$$
$$= argmax_{H_i,1\leq i\leq n}P(H_i,O_i)$$
$$= argmax_{H_i,1\leq i\leq n-1}P(O_i|H_i)P(H_i|H_{i-1})P(H_{i-1}|O_{i-1}).$$

Thus if $p_i = P(H_i|O_i)$, then

$$p_i = H^* = argmax_{H_i,1\leq i\leq n} \prod_{i=2}^{n-1} P(O_i|H_i) \prod_{i=2}^{n} P(H_i|H_{i-1})p_1.$$

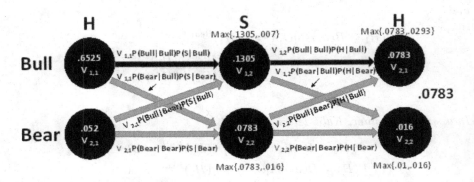

FIGURE 2.5
Example of recursive decisions in a Viterbi Algorithm matrix.

Algorithm 3. *Viterbi Algorithm*

Given $O_1, O_2, \ldots, O_n$ are the observed random variables and $H_1, H_2, \ldots, H_n$ are the hidden random variables with possible outcomes $Y_1, Y_2, \ldots, Y_{n_h}$.

1. Create a $n_h \times n$ Viterbi Algorithm Matrix, V, where the entries

$$V_{i,j} = \begin{cases} P(O_j|Y_i)P(Y_i), & \text{if } j = 1, \\ max_{1 \leq k \leq n_h} V_{k,j-1}P(Y_i|Y_k)P(O_j|Y_i), & \text{if } 1 < j \leq n. \end{cases}$$

2. Determine $argmax_{H_i,1\leq i\leq n}P(H_i,O_i) = max_{1\leq j\leq n_h}V_{i,j}$.

3. Recall that $argmax_{H_i,1\leq i\leq n}P(H_i|O_i) = \frac{argmax_{H_i,1\leq i\leq n}P(H_i,O_i)}{P(O)}$.

4. Backtrack through the Viterbi Algorithm matrix determining the largest probability in each column, if $V_{m,n}$ is the largest value in column j then the sequence determined with $H_n = Y_m$ will provide maximum likelihood of the observation sequence, O, occurring.

Example 2.14. *Kalu knows that Rath is {Happy, Sad, Happy} on three consecutive days and he wants to know the most likely hidden states of the stockmarket on those days.*

We will begin by finding the Viterbi Algorithm matrix. The recursive ideas behind each entry in the Viterbi Algorithm matrix can be seen in Figure 2.5.

$$V = \begin{matrix} \textit{Bullish} \\ \textit{Bearish} \end{matrix} \begin{pmatrix} H & S & H \\ .6525 & .1305 & .0783 \\ .052 & .0783 & .016 \end{pmatrix}.$$

If M is the hidden sequence that maximizes the probability of the observed sequence occurring, then

$$P(M, \{Happy,\ Sad,\ Happy\}) = \max\{.0783, .016\} = .0783$$

and the maximum likelihood of

$$P(M|\{Happy,\ Sad,\ Happy\}) = \frac{.0783}{P(\{Happy,\ Sad\ Happy\})}$$
$$= \frac{.0783}{.67 \cdot .345 \cdot .65} \approx .52$$

where $P(\{Happy,\ Sad,\ Happy\})$ can be determined using the Forward Algorithm in a similar manner to Example 2.11.

Kalu really wishes to find the hidden sequence that maximizes this likelihood. Revisiting the Viterbi Algorithm matrix and the recursive paths in Figure 2.5, choose the hidden states with the largest probabilities working backwards. Thus the most likely hidden sequence M is

$$\{Bullish,\ Bullish,\ Bullish\}.$$

The Viterbi Algorithm is a widely used algorithm. For those of you who use speech or hand writing recognition on your personal devices, the Viterbi Algorithm may be used to decipher what you are trying to communicate.

2.2.3 The Learning Problem

In both the likelihood and the decoding problems, the model was assumed and fixed. That is, we knew the initial probabilities, π, the transmission matrix, T, and emission matrix, E.

This final type of problem, the learning problem, requires an algorithm that will help determine the optimal model $\lambda = (T, E, \pi)$.

In this section, we will discuss the **Baum-Welch Algorithm**.

Given an initial sequence of observed states $O = (O_1, O_2, \ldots, O_n)$, called the **training sequence**, and a hidden sequence $H = (H_1, H_2, \ldots, H_n)$, the Baum-Welch Algorithm will find another hidden sequence H' such that

$$P(O|H') \geq P(O|H).$$

Ideally, we wish to find H' such that $P(O|H') \geq P(O|H)$ for all H, however, that is not guaranteed under the Baum-Welch Algorithm.

Algorithm 4. *(Baum-Welch Algorithm)*

The Baum-Welch Algorithm is an iterative algorithm using the following iterative steps, where α, β, and γ are defined in the Forward/Backward Algorithm.

1. *Interpolate out one step by finding the probability,*

$$\xi_{i,j}^k = P(H_k = j, H_{k+1} = i | O, \lambda)$$

$$= \frac{\beta(H_{k+1})^i E_{(O_{k+1},i)} T_{i,j} \alpha(H_k)^j}{P(O|\lambda)}.$$

$$= \frac{\beta(H_{k+1})^i E_{(O_{k+1},i)} T_{i,j} \alpha(H_k)^j}{\sum_{i=1}^{n-1} \sum_{j=1}^{n-1} \beta(H_{k+1})^i E_{(O_{k+1},i)} T_{i,j} \alpha(H_k)^j}.$$

It is important to note that the Forward algorithm, determining $\alpha's$, is about determining pre-priori information and the Backward algorithm is about smoothing the information, determining $\beta's$, the post-priori information.

A visualization of this idea can be seen in Figure 2.6. In order to move from $\alpha(H_j)^t$ to $\beta(H_i)^{t+1}$, we need to move from H_j to H_i, thus multiplying by $T_{i,j}$. Then we have to think about how we move to $\beta(H_i)^{t+1}$ with observation state O_{t+1}. Thus,

$$\xi_{i,j}^t = \beta(H_i)^{t+1} E_{(O_{t+1},i)} T_{i,j} \alpha(H_j)^t.$$

Since we are calculating a probability, we also have to divide by a normalizing factor, $P(O|\lambda)$.

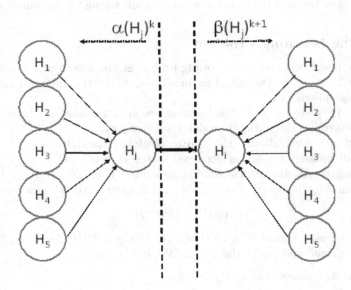

FIGURE 2.6
Visualizing the Role of α and β.

2. *Define a new model with* $\pi^{new} = \gamma(H_0)$

$$T_{i,j}^{new} = \frac{\# \text{ of transitions from state } H_j \text{ to state } H_i}{\# \text{ of transitions from state } H_j}$$

$$= \frac{\sum_{k=1}^{n-1} \xi^k(i,j)}{\sum_{k=1}^{n-1} \gamma(H_j)^k}$$

$$E_{i,j}^{new} = \frac{\# \text{ of times the hidden state is } H_j \text{ given the observation is } O_i}{\# \text{ of transitions from state } H_j}$$

$$= \frac{\sum_{k=1,O_k=i}^{n} \gamma(H_j)^k}{\sum_{k=1}^{n-1} \gamma(H_j)^k}$$

Note that

$$\sum_{t=1}^{T-1} \gamma_t(i)$$

is the expected number of transitions from H_i and

$$\sum_{t=1}^{T-1} \xi^t(i)$$

is the expected number of transitions from H_i to H_j.

Let's put the Baum-Welch Algorithm in action through an example.

Example 2.15. *Scientists observed a short nucleotide sequence O=AGCT with model λ defined by the transition, emission matrix*

$$T = E = \begin{pmatrix} & A & C & G & T \\ A & .3 & .322 & .248 & .177 \\ C & .205 & .298 & .246 & .239 \\ G & .285 & .078 & .298 & .292 \\ T & .210 & .302 & .208 & .292 \end{pmatrix},$$

and $\pi = \{.2, .3, .3, .2\}$.

Unfortunately, it is common to see actual genetic makeup, and the scientist wishes to determine if in fact a CG sequence is present in the original genetic makeup.

We use Forward/Backward Algorithm to determine

$$\alpha(H_1) = \{0.06, 0.0966, 0.0744, 0.0354\}$$
$$\alpha(H_2) = \{0.021, 0.0053, 0.017, 0.0197\}$$
$$\alpha(H_3) = \{0.0032, 0.0044, 0.0042, 0.0037\}$$
$$\alpha(H_4) = \{0.0009, 0.0012, 0.0007, 0.0012\}$$
$$\beta(H_4) = \{1,1,1,1\}$$
$$\beta(H_3) = \{0.2455, 0.262, 0.2491, 0.2553\}$$
$$\beta(H_2) = \{0.0543, 0.0747, 0.0588, 0.0614\}$$
$$\beta(H_1) = \{0.0188, 0.0049, 0.0169, 0.0187\}$$
$$\gamma(H_1) = \{0.32, 0.134, 0.357, 0.188\}$$
$$\gamma(H_2) = \{0.304, 0.106, 0.267, 0.323\}$$
$$\gamma(H_3) = \{0.2, 0.293, 0.266, 0.24\}$$
$$\gamma(H_4) = \{0.225, 0.3, 0.175, 0.3\}$$

$$\xi^1 = \begin{pmatrix} 0.121 & 0.21 & 0.125 & 0.042 \\ 0.031 & 0.073 & 0.047 & 0.021 \\ 0.131 & 0.058 & 0.169 & 0.079 \\ 0.099 & 0.228 & 0.121 & 0.081 \end{pmatrix}, \xi^2 = \begin{pmatrix} 0.152 & 0.041 & 0.102 & 0.084 \\ 0.162 & 0.059 & 0.157 & 0.177 \\ 0.176 & 0.012 & 0.149 & 0.169 \\ 0.129 & 0.047 & 0.104 & 0.169 \end{pmatrix},$$

$$\xi^3 = \begin{pmatrix} 0.096 & 0.142 & 0.104 & 0.066 \\ 0.094 & 0.189 & 0.149 & 0.127 \\ 0.09 & 0.034 & 0.124 & 0.107 \\ 0.093 & 0.185 & 0.122 & 0.15 \end{pmatrix}, \sum_{t=1}^{3} \xi^t = \begin{pmatrix} 0.096 & 0.142 & 0.104 & 0.066 \\ 0.094 & 0.189 & 0.149 & 0.127 \\ 0.09 & 0.034 & 0.124 & 0.107 \\ 0.093 & 0.185 & 0.122 & 0.15 \end{pmatrix}.$$

We can use this information to determine that the expected number of transitions from state C to state G is

$$\sum_{t=1}^{3} \xi^t(2,3) = .321.$$

Example 2.16. *Perhaps Kalu wishes to better his model to determine the behavior of the stock market based on Rath's reported mood. In Examples*

2.11 through 2.13, we determined α, β, and γ using the Forward/Backward Algorithm.

$$\xi^1 = \begin{pmatrix} 0.607871 & 0.0302771 \\ 0.301736 & 0.060116 \end{pmatrix}$$

$$\xi^2 = \begin{pmatrix} 0.569295 & 0.243872 \\ 0.0688528 & 0.11798 \end{pmatrix}$$

$$\xi^3 = \begin{pmatrix} 0.69083 & 0.109367 \\ 0.122334 & 0.0774684 \end{pmatrix}$$

$$\xi^4 = \begin{pmatrix} 0.50013 & 0.0587622 \\ 0.300078 & 0.141029 \end{pmatrix}.$$

From here we can construct an updated model with

$$\pi^{new} = \gamma(H_1) = \{.91, .09\},$$

$$T^{new} = \begin{pmatrix} 0.74914 & 0.52723 \\ 0.25086 & 0.47277 \end{pmatrix}, E^{new} = \begin{pmatrix} 0.678328 & 0.3725 \\ 0.321672 & 0.6275 \end{pmatrix}$$

This is just one iteration of the Baum-Welch Algorithm. In order to progress toward an even better model, the Baum-Welch Algorithm should be iterated multiple times.

Baum-Welch Algorithm is just one method in a larger category of **Expectation Maximization algorithms** that are generally used in machine learning to produce advanced models. There are a variety of **gradient** methods like Baum-Welch Algorithm that have been developed with similar goals in mind.

2.3 CASE STUDY: Spread of Infectious Disease

In 2019, the world broke out into an international pandemic which led many to start to think more closely about the mathematics behind the spread of infectious diseases. Many diseases, like COVID-19, in simple terms would be modeled with SIR models, with humans in three categories, S (Susceptible), I (Infectious), and R (Recovered). There are more complex SIR models that provide for temporary immunity, quarantine, or even death.

We will explore a variety of these in this case study. In addition, most infectious disease models would implement a stochastic continuous-time Markov chain model; however, we will focus on discrete time models in this particular case study.

We begin with the basic SIR model. This model has the underlying assumption that birth and death occur at roughly the same rate throughout time. We will also assume that the time step in this model is days; however, parameters could be adjusted to accommodate for other units of time.

If the states of the Markov chain are X_1, X_2, ..., and

$$P(X_t = R | X_{t-1} = R) = 1$$

then recovery is an absorbing state which would happen in a case when immunity is guaranteed. As of May 2020, there were hopes that some temporary immunity existed but evidence from the outbreak in China showed that if it did exist it was fairly short lived.

We begin by assuming that social distancing is not in place and that there may be some partial immunity based on antibodies but that it is not guaranteed.

One example of a transition matrix,

$$T = \begin{pmatrix} S & I & R \\ 0.99998 & 0. & 0.0714286 \\ 0.00002 & 0.998 & 0. \\ 0. & 0.002 & 0.928571 \end{pmatrix}.$$

produces a spread of infection shown in Figure 2.7.

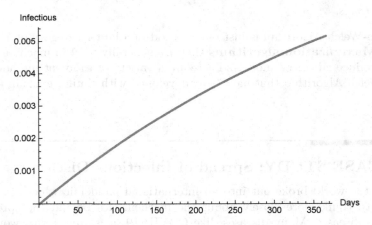

FIGURE 2.7
Growth of disease under a simple discrete SIR model.

The **reproduction number**, R_0, is the expected number of secondary cases that one case would produce in an entire susceptible population.

A positive reproduction number is indicative of a possible probability of outbreak, while a negative R_0 indicates that an outbreak is not possible.

$$R_0 = \frac{\text{infection rate}}{\text{the expected disease length}}.$$

In the simplest models, R_0 can be estimated by the rate of infection, r_0, and the estimated length of the infection, L, where

$$R_0 \approx 1 + r_0 L.$$

Thus for the model shown in Figure 2.7, if we estimate $r_0 \approx .017$ and $L = 21$ days then $R_0 \approx 1.357$.

If instead, immunity is present and thus recovery acts as an absorbing state, one might see a behavior such as in Figure 2.8.

In this case, let's assume that the transition matrix is

$$T = \begin{pmatrix} & S & I & R \\ & 0.98 & 0 & 0 \\ & 0.02 & 0.96 & 0 \\ & 0 & 0.04 & 1 \end{pmatrix}.$$

Then the the expected time from susceptible to recovery can be found using

$$N = (I - Q)^{-1} = \begin{pmatrix} 50 & 0. \\ 25 & 25 \end{pmatrix}.$$

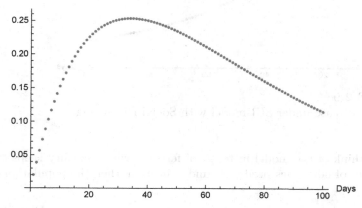

FIGURE 2.8
Growth of disease under SIR model with immunity.

Thus we should expect that a susceptible human will stay susceptible for 50 days and then stay infectious for 25 days before reaching a permanent recovery.

Let's expand our model of the epidemic to include one additional state, susceptible and social distancing (SS). We will assume that there are still susceptible individuals who are not social distancing (S); however, for those who social distance, the transmission to other individuals is much less likely, and those who are recovered make the decision to social distance in the future.

Using the transition matrix

$$T = \begin{pmatrix} S & SS & I & R \\ 0.9 & 0. & 0. & 0 \\ 0. & 0.99 & 0 & 0. \\ 0.1 & 0.01 & 0.96 & 0. \\ 0. & 0. & 0.04 & 1 \end{pmatrix}$$

the behavior can be seen in Figure 2.9. In this model, a susceptible person who is not social distancing, will become infected within 10 days on average, where a susceptible person who is social distancing is expected to take about 100 days to get infected with the disease.

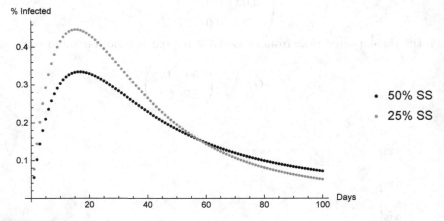

FIGURE 2.9
Growth of disease under SIR model with Social Distancing.

If we think of the model in terms of fertility, where fertility keeps track of the number of new cases produced, and transition, then the population can be described

$$P = F + T,$$

where F is a matrix representing the fertility among classes and T represents the transition between classes. If $(I - T)^{-1}$ exists, we can determine the net reproductive value, n, [7]

$$n = \rho((I - T)^{-1} F)$$

and the exponential growth rate, $\rho(P)$.

Example 2.17. *In this disease model, we are thinking of the offspring as those in the infectious category, so we reorder our rows and columns so that the first row/column is the infectious category. Let*

$$T = \begin{pmatrix} & I & S & R \\ I & 0.96 & 0 & 0 \\ S & 0.04 & 1 - 0.0714286 & 0 \\ R & 0 & 0.0714286 & 0.99998 \end{pmatrix} \quad and \quad F = \begin{pmatrix} 0.4 & 0 & 0.00002 \\ 0 & 0 & 0 \\ 0 & 0 & 0 \end{pmatrix}.$$

$P = F + T$ *and the net reproductive value* $n = 10$ *and the exponential growth rate* $\rho(P) = 1.36$.

2.4   CASE STUDY: Text Analysis and Autocorrect

In Chapter 3, we will explore data compression techniques in order to do comparative analyses of text. Here we explore different ways that Markov chains can be applied to text analysis by looking at a small text analysis of Jane Austen's work.

There are over 13,000 unique word pairings in Jane Austen's texts, we explore just a few with the transition matrix.

$T =$

	miss	anne	poor	marianne	when	replied	with	felt	jane	appeared	cried	but
miss	0	0	0.04	0	0.10	0	0	0	0.08	0	0.05	0.04
anne	0.93	0	0.64	0	0	0.58	0	0	0	0	0.75	0.7
poor	0	0	0	0	0.10	0	0	0	0	0	0	0.04
marianne	0.02	0	0.04	0	0.10	0.03	0.01	0.13	0	0	0.10	0.04
when	0.01	0.09	0	0.04	0	0	0.01	0.13	0.08	0.2	0.05	0.04
replied	0	0.09	0	0.04	0.00	0	0	0	0	0	0	0
with	0	0.09	0.04	0.04	0.10	0.03	0	0.13	0	0.2	0.05	0.04
felt	0	0.18	0	0.08	0	0	0	0	0.15	0	0	0.06
jane	0.03	0	0.04	0.00	0.10	0.06	0.01	0	0	0	0	0.04
appeared	0	0	0	0.46	0	0	0	0	0	0	0	0
cried	0	0.18	0	0.13	0	0	0	0	0.23	0	0	0
but	0	0.36	0.21	0.21	0.50	0.29	0.07	0.63	0.46	0.6	0	0

For this small study, we will assume that the words in the transition matrix are the only ones present in Austen's work. Python and R syntax to find a database of word pairs to create a similar transition matrix for all of Austen's work similar can be found at Github links 11 and 12.

You should not be surprised to hear that a matrix with so many word choices as that which is actually in these works would produce a very **sparse matrix** with many entries equal to zero.

If we wished to create a random quote of 4 words from Jane Austen's likely transitions in T and we wanted the quote to start with *Marianne*, we might look at the adjacencies to T_4 since *Marianne* is represented in the 4th row of T.

$$\{I,\ T,\ T^2,\ T^3\}.\{0,0,0,1,0,0,0,0,0,0,0,0\}.$$

	I	T	T^2	T^3
miss	0	0	0.02	0.03
Anne	0	0	0.26	0.3
poor	0	0	0.01	0.03
marianne	1	0	0.04	0.05
when	0	0.04	0.12	0.06
replied	0	0.04	0	0.03
with	0	0.04	0.12	0.06
felt	0	0.08	0.01	0.08
jane	0	0	0.02	0.04
appeared	0	0.46	0	0.02
cried	0	0.12	0	0.06
but	0	0.21	0.39	0.26

(with *Marianne* heading the columns),

seen in the columns of the matrix above. To generate our sentence, we may choose to run a simulation using these probabilities to choose our word sequence or choose the most likely sentence to occur based on these probabilities,

Marianne appeared but Anne.

In more practical terms, as mentioned earlier in this chapter, hidden Markov models are regularly used in spoken and handwritten word recognition. Smartphones have built-in software to both suggest words and autocorrect words as you are typing a text.

We will take a deeper look into how hidden Markov chains can be applied to autocorrect or autocomplete in this case study. Suppose that the observed word set, the words that are in the text, that we wish to look at is

$$\{heal,\ hear,\ here,\ heel,\ hele,\ Hela\}$$

The language-correct software will choose the word that is most likely to occur, but may not choose until there is a dominant choice. For example, let's assume that the first two letters in the word are known to be *he* and that the transition matrix between letters is known to be

$$
T = \begin{array}{c} a \\ e \\ l \\ r \end{array}
\left(
\begin{array}{cccc}
a & e & l & r \\
.10 & .29 & .33 & .21 \\
.17 & .18 & .40 & .36 \\
.28 & .21 & .27 & .29 \\
.45 & .32 & 0 & .14
\end{array}
\right).
$$

Like the Jane Austen example presented earlier, given that the first two letters in the word are *he* we will look at $\{T,\ T^2\}.\{0,1,0,0\}$, presented in the columns in the matrix below, to determine the most likely choice for the last two letters in the word,

$$
\begin{array}{c} a \\ e \\ l \\ r \end{array}
\left(
\begin{array}{cc}
0.29 & 0.22 \\
0.18 & 0.28 \\
0.21 & 0.27 \\
0.32 & 0.23
\end{array}
\right),
$$

which is *here*.

Next, let's assume that you have typed a three letter word incorrect as *aer*. Your phone software determines that this word is incorrect since it is not in it's dictionary database, so the software chooses to autocorrect.

Your typed word provides the observation sequence

$$O = \{a,\ e,\ r\}.$$

and we wish to determine the most likely hidden sequence. In this case, we will assume that the only possible hidden states are a, e, l, and r.

We assume that the initial probability is $\pi = \left\{\frac{3}{4},\ \frac{1}{8},\ 0,\ \frac{1}{8}\right\}$.

Using the Viterbi Algorithm, the most likely hidden sequence is *are*, which can be seen in Figure 2.10. The likelihood of the hidden sequence *are* in this situation is approximately .06%.

In order to access the bigram data used in this study, go to Github link 13.

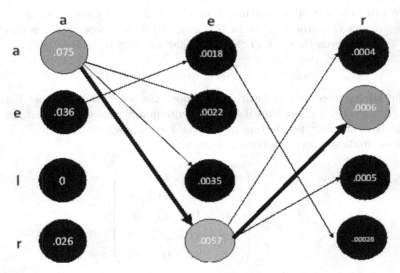

FIGURE 2.10
Viterbi Algorithm Example for Autocorrect.

2.5 CASE STUDY: Tweets and Time Series

In [16] and [25], the authors discuss how hidden Markov models can be used to predict stock market behavior. In this case study, we take a close look at a few different observation sequences for making such predictions.

We begin with an observation sequence described by states related to data on the S&P 500 index. Continuous Markov chains and continuous HMMs can be applied in this situation. Since we have not focused on continuous models, an altered discrete model is discussed here.

In this case study, an observation state, O_t, will be defined as a three-dimensional binary vector where

$$
O_t = \left(O_t^{(1)}, O_t^{(2)}, O_t^{(3)} \right)
$$
$$
= \left(1 + \left\lfloor \frac{close - open}{open} \right\rfloor, 1 + \left\lfloor \frac{high - open}{open} \right\rfloor, 1 + \left\lfloor \frac{low - open}{open} \right\rfloor \right),
$$

where open refers to the opening value of the S&P 500 index on day t, closing is the day t closing value, and high and low are the high and low values of day t.

Note that a trader is looking for increasing and decreasing behaviors and a 1 in O_t would represent an increase, where a 0 would represent a decrease. These particular observation states will allow for a record of variation throughout time.

Since we wish to use trends to predict future values, we will assume a $d = 2$ day latency period. For simplicity, d is chosen to be small for this case study. Longer latency period could be selected and analyzed in a similar manner.

Thus a trader's problem becomes the following:

Given the observation sequence $O = (O_1, O_2)$ and the model λ, predict general relationships between opening, closing, low, and high values on day 3.

We will use the forward-backward method to determine the hidden sequence $H = (O_2, O_3)$. More particularly, we wish to determine

$$H^* = argmax_{O_3} P(O_3|O, \lambda)$$
$$= argmax_{O_3} P(O_1, O_2, O_3|\lambda).$$

In this case study, we demonstrate with a small training set pulled from [13] from May 2019, which can be seen in Table 2.1. A larger data set for further study can be found at Github link 16.

TABLE 2.1
Observation State, O, in May 2019 for S&P 500.

Date, t	5/1	5/2	5/3	5/6	5/7	5/8	5/9
$(O_t^{(1)}, O_t^{(2)}, O_t^{(3)})$	(0,1,0)	(0,1,0)	(1,1,1)	(1,1,0)	(0,1,0)	(0,1,0)	(1,1,0)
Date, t	5/10	5/13	5/14	5/15	5/16	5/17	5/20
$(O_t^{(1)}, O_t^{(2)}, O_t^{(3)})$	(1,1,0)	(0,1,0)	(1,1,1)	(1,1,0)	(1,1,1)	(1,1,0)	(0,1,0)
Date, t	5/21	5/22	5/23	5/24	5/28	5/29	5/30
$(O_t^{(1)}, O_t^{(2)}, O_t^{(3)})$	(1,1,1)	(1,1,0)	(0,1,0)	(0,1,0)	(0,1,0)	(0,1,0)	(1,1,0)

The probabilities for O_t, listed in Table 2.2, are found using the data in Table 2.1. Keep in mind that there are eight possible triplets for O_t; however, only the states that were prevalent in the May 2019 (Table 2.1) data are presented here.

TABLE 2.2
$P(O)$.

O	(0,1,0)	(1,1,0)	(1,1,1)
π	.524	.286	.190

The transmission matrix

$$T = \begin{pmatrix} & (0,1,0) & (1,1,0) & (1,1,1) \\ (0,1,0) & .454 & .66 & 0 \\ (1,1,0) & .273 & .17 & 1 \\ (1,1,1) & .273 & .17 & 0 \end{pmatrix}.$$

Using this transmission matrix,

$$T^2 = \begin{pmatrix} & (0,1,0) & (1,1,0) & (1,1,1) \\ (0,1,0) & 0.386 & 0.412 & 0.66 \\ (1,1,0) & 0.443 & 0.379 & 0.17 \\ (1,1,1) & 0.17 & 0.209 & 0.17 \end{pmatrix},$$

If the trader observes that the $O_{5/29} = (0,1,0)$ but does not know the observation of $O_{5/30}$, then using T^2, the trader might guess that the behavior on May 31st would be (1,1,0). This means that the closing and high values are larger than the opening, and that the low value is lower than the opening value on that day.

In addition, the distribution $P(O_3)$ can be determined

$$T^2\pi = P(O_3) = \begin{cases} .446 & \text{if } O_3 = (0,1,0), \\ .373 & \text{if } O_3 = (1,1,0), \\ .181 & \text{if } O_3 = (1,1,1), \\ 0 & \text{otherwise.} \end{cases}$$

This tells the trader that the most probable observation given during the month of data is (0,1,0). The trader could use this information to simulate the behavior of the closing cost relative to the opening over a future period of time as well.

There are many ways that investors predict the future of their products, including stocks. Another way to do this is through sentiment analysis of tweets, reviews, and other types of posts. Similar to Rath and Kalu's analysis presented earlier in this chapter, studies have shown that the market can react to media statements from prominent leaders, such as presidential tweets [21].

If a trader wishes to study the market from this perspective, they could scrape tweets from a variety of sights, such as that seen in [30], and set up a word bank to search for words that resemble bearish or bullish market behavior.

For the following example, we pull 11 Trump tweets from roughly the beginning of 11 consecutive months, shown in Table 2.3. Words that should be considered in a sentiment word bank are italicized. In addition, the opening of the S&P 500 index on each of the day of the tweet and the following trading day are shown in Table 2.4.

A larger dataset of Trump tweets about the stock market can be found at Github link 18.

TABLE 2.3

Date	Tweet	Sentiment
10/2/2019	"All of this *impeachment* nonsense, which is going nowhere, is driving the Stock Market, and your 401K's, *down*."	Bearish
11/1/2019	"Stock Market *up* BIG!"	Bullish
12/2/2019	"U.S. Markets are *up* as much as 21% since the announcement of Tariffs"	Bullish
1/9/2020	"STOCK MARKET AT ALL-TIME *HIGH*!	Bullish
2/4/2020	"Market *up* big today on very *good* economic news	Bullish
3/9/2020	"...That, and the Fake News, is the reason for the market *drop*!"	Bearish
4/8/2020	"... Our Economy will *BOOM*, perhaps like never before!"	Bullish
5/13/2020	"When the so-called rich guys speak *negatively* about the market, you must always remember that some are betting big *against* it...."	Bearish
6/5/2020	"It's a *stupendous* number. It's joyous, let's call it like it is."	Bullish
7/21/2020	"Vote for the Radical Left with their BIG *Tax Hikes*"	Bearish
8/24/2020	"Joe would end it all and close it all *down*."	Bearish

TABLE 2.4
S&P 500 index behavior trends.

Date	Opening	Date	Opening	Trend
10/2/2019	2924.78	10/3/2019	2885.37	D
11/1/2019	3050.72	11/4/2019	3078.95	U
12/2/2019	3143.85	12/3/2019	3087.40	D
1/9/2020	3266.03	1/10/2019	3281.81	U
2/4/2020	3280.61	2/5/2020	3324.90	U
3/9/2020	2863.89	3/10/2020	2813.47	D
4/8/2020	2685	4/9/2020	2776.98	U
5/13/2020	2865.86	5/14/2020	2794.54	D
6/5/2020	3163.84	6/8/2020	3199.92	U
7/21/2020	3268.52	7/22/2020	3254.86	D
8/24/2020	3418.09	8/25/2020	3435.95	U

In this example, we will use the sentiment of Trump tweets in order to predict the change from opening to closing behavior in the S&P 500 index. We will assume that stock behaviors are monthly; however, in a real-life model with a much larger data set, a similar analysis should be done with daily trends. The data from Table 2.3 is used to create the transmission matrix

$$T = \begin{matrix} \\ Up \\ Down \end{matrix} \begin{pmatrix} Up & Down \\ \frac{1}{5} & 1 \\ \frac{4}{5} & 0 \end{pmatrix}.$$

and the emission matrix

$$E = \begin{matrix} \\ Bull \\ Bear \end{matrix} \begin{pmatrix} U & D \\ \frac{5}{6} & \frac{1}{5} \\ \frac{1}{6} & \frac{4}{5} \end{pmatrix}.$$

In addition,

$$\pi = (P(Up),\, P(Down)) = \left(\frac{6}{11}, \frac{5}{11} \right).$$

We begin by studying three month, or one quarter, trends and finding the most likely observation sequence

$$O = (O_1, O_2, O_3),$$

where O_i represents the sentiment in Trump's tweet at time i in the three month sequence and

$$H = (H_1, H_2, H_3),$$

where H_i represents the difference between closing and opening on the ith month of the three month sequence. Notice that

$$E \cdot T \cdot \pi = (0.55697, 0.44303) \text{ and } E \cdot T^2 \cdot \pi = (0.547758, 0.452242)$$

and thus the most likely scenario in that Trump's three month tweet sentiment will be

(Bullish, Bullish, Bullish).

Additionally, the unit eigenvector, v, associated with λ_{PF}, for $E.T$,

$$v = (0.553097, 0.446903).$$

Thus, over a long period of time, it is much more likely that the monthly sentiment for Trump's tweets is Bullish, based on the given data.

More importantly, one might wish to know, if Trump's monthly tweets are of a certain observation sequence, such as (Bullish, Bullish, Bullish), what is happening in the market?

In this case Viterbi's Algorithm can be followed. A diagram of Viterbi's Algorithm can be found in Figure 2.11 which shows that if Trump's three month

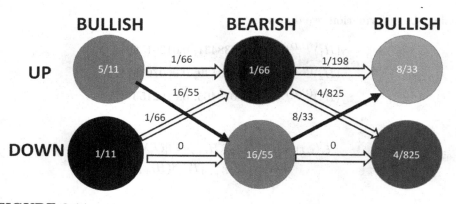

FIGURE 2.11
Viterbi's Algorithm for three month marker trends based on tweets.

tweet sentiment is (Bullish, Bearish, Bullish) the most likely hidden sequence of the market is

(Up, Down, Up) with probability .2523.

Finally, we wish to find the best model. We begin by assuming that

$$O = (Bullish, Bullish, Bullish),$$

the first three states in Table 2.3, then

$$\alpha(H_1)^U = \frac{5}{11}, \quad \alpha(H_1)^D = \frac{1}{11},$$
$$\alpha(H_2)^U = \frac{1}{33}, \quad \alpha(H_2)^D = \frac{16}{55},$$
$$\alpha(H_3)^U = \frac{49}{198}, \quad \alpha(H_3)^D = \frac{4}{825},$$

$$\beta(H_3)^U = 1, \quad \beta(H_3)^D = 1,$$
$$\beta(H_2)^U = \frac{9}{150}, \quad \beta(H_2)^D = \frac{5}{6},$$
$$\beta(H_1)^U = \frac{2449}{4500}, \quad \beta(H_1)^D = \frac{49}{900}.$$

Note that these values tell us that

$$P(O|\lambda) = \sum_k \alpha(H_3)^k = 0.2523.$$

Using this information, we determine

$$\gamma(H_1)^{(U,D)} = \{0.98038431, 0.01961569\},$$
$$\gamma(H_2)^{(U,D)} = \{0.03923139, 0.96076861\},$$
$$\gamma(H_3)^{(U,D)} = \{0.98078463, 0.01921537\},$$

$$\xi_t(i,j) = P(H_t = j, H_{t+1} = i|O,\lambda)$$
$$= \alpha(H_i)E_{H_{t+1},j}T_{i,j}\beta(H_{t+1})$$

and the new model is $\pi^{new} = \{0.98038431, 0.01961569\}$.

$$T^{new} = \begin{pmatrix} 0.51020408 & 1 \\ 0.48979592 & 0 \end{pmatrix}, E^{new} = \begin{pmatrix} 0.96153846 & 0.01960784 \\ 0.03846154 & 0.98039216 \end{pmatrix}.$$

Keep in mind that this is just one iteration of the Baum-Welch algorithm. Python and R code related to this case study can be found at Github links 14 and 15, respectively.

2.6 Exercises

1. Determine which of the following matrices are primitive or irreducible.
$$A_1 = \begin{pmatrix} 0 & 2 \\ 1 & 1 \end{pmatrix}, A_2 = \begin{pmatrix} 2 & 0 \\ 1 & 1 \end{pmatrix}$$

2. Assume that from the beginning of time, UNC, Duke, and NC State admitted only male students. Also assume that, at that time, 70 percent of the sons of UNC men went to UNC and the rest went to NC State, 50 percent of the sons of NC State men went to NC State, and the rest split evenly between UNC and Duke; and of the sons of Duke men, 80 percent went to Duke, 10 percent to UNC, and 10 percent to NC State.

   If this were still the case (and these were the only 3 universities), use the Perron-Frobenius Theorem to determine what percent of students would be at each institution long into the future.

3. If the students that attended Duke live in Raleigh, Chapel Hill, and Durham with probabilities 25%, 35%, and 40% respectively. While students that attend UNC live in Raleigh, Chapel Hill, and Durham with probabilities 30%, 60%, and 10%. Finally, the students that attend NC State live in Raleigh, Chapel Hill and Durham with probability 60%, 20%, and 20%. Those who are from Durham, Raleigh, and Chapel Hill, go to Duke (50%), UNC (20%), and NC State (30%).

   Determine the probability that a student goes to Duke given that they are from Durham.

4. An epidemic is beginning to become prevalent in two different countries, A and B. Citizens of those countries can be in one of three states, susceptible (S), infectious (I), or recovered (R). From our daily observations a few weeks into the outbreak, we note that the probabilities between the observed states of the infection among its citizens and the probabilities between the states of the infection.

If the population of country A is twice as large as the population of country B, and we randomly sample four citizens from the combined countries, use hidden markov chains to determine the probability that they are from countries AABB given that their status is SIRS.

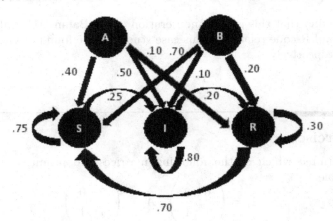

5. After thinking about Example 2.14, determine the following
   a. $P(O_1 = Sad)$
   b. $P(O_1 = Sad, O_2 = Sad)$
   c. $P(O_1 = Sad, O_2 = Sad, O_3 = Happy)$
   d. The maximum likelihood of the observation sequence $\{Sad, Sad, Happy\}$ occurring.
   e. The most likely hidden sequence given that Rath is $\{Sad, Sad, Happy\}$.

6. In the theme of attempting to find the best model, use the information from Example 2.16, apply the Baum-Welch Algorithm one additional time in order to find a new model $\lambda = (\pi, T, E)$.

7. Doctors are trying to determine the original genetic making up of their patients given mutated DNA sequences collected from biopsied patients. You are given a mutated DNA sequence, **ATG** where, A (adenine), C (cytosine), G (guanine), and T (thymine). The transmission matrix, T, and emissions matrix, E, are as follows.

$$
T = E = \begin{pmatrix}
 & A & C & G & T \\
A & .40 & .225 & .400 & .252 \\
C & .20 & .325 & .100 & .223 \\
G & .15 & .300 & .200 & .177 \\
T & .25 & .150 & .300 & .348
\end{pmatrix}
$$

In addition, assume that the four state, A, C, G, and T, are equally likely. Determine the likelihood of the observation sequence **ATG** using the forward algorithm.

8. Using the information from Exercise 7 and the Forward/Backward algorithm, find the most likely original molecular structure of the genetic code with the four possible hidden state, A (adenine), C (cytosine), G (guanine), and T (thymine).

3

SVD and PCA

Like Chapter 1, this chapter on clustering techniques and their application to data science. Chapter 1 relies heavily on powers of matrices, eigenvalues, and eigenvectors, which are only relevant for square matrices. This chapter digs deeper into techniques that are more practical in terms, not imposing the limitation that matrices must be square.

We begin with an overview of vector and inner product spaces since orthogonality is pertinent in the discussion of further topics in the chapter.

3.1 Vector and Inner Product Spaces

A set of vectors, V, under defined vector addition and scalar multiplication is called a **vector space** if the following properties hold.

1. V is closed under vector addition.

2. V is closed under scalar multiplication.

3. For all u and v in V, $u + v = v + u$.

4. For all u,v and w in V,

$$(u + v) + w = u + (v + w).$$

5. There exists an unique additive identity in V.

6. Every vector $u \in V$ has an additive inverse $-u \in V$.

7. For each vector $u \in V$, $1 \cdot u = u$.

8. For all u and v in V and scalar k, $k(u + v) = ku + kv$.

9. For all u in V and scalars k and l, $(k + l)u = ku + lu$.

10. For all u and v in V and scalar k, $k(lu) = (kl)u$.

Example 3.1. *The set of 2×2 symmetric matrices under standard matrix addition and scalar multiplication is a vector space; however, the set of 2×2 invertible matrices under standard matrix addition and scalar multiplication is not.*

DOI: 10.1201/9781003025672-3

Notice that the set of 2×2 *invertible matrices is not closed under matrix addition. An example of two* 2×2 *invertible matrices whose sum is not invertible is*

$$\begin{pmatrix} 1 & 0 \\ 0 & 1 \end{pmatrix} + \begin{pmatrix} -1 & 0 \\ 0 & -1 \end{pmatrix} = \begin{pmatrix} 0 & 0 \\ 0 & 0 \end{pmatrix}.$$

An **inner product** on a set V is a function that maps order pairs $(x,y) \in V \times V$ to a number $< x,y >$ while satisfying the following properties:

1. For all $v \in V$, $< v,v > \geq 0$ and $< v,v >= 0$ if and only if $v = \vec{0}$.
2. For all u,v, and $w \in V$, $< u,v + w >=< u,v > + < u,w >$.
3. For all $u,v \in V$ and scalars k, $< ku,v >=< u,kv >= k < u,v >$.
4. For all $u,v \in V$, $< u,v >= \overline{< v,u >}$.

A vector space with a defined inner project is called an **inner product space**.

In an inner product space, V, a set of vectors is called **orthogonal** if for each pair of vectors v and w in V,

$$< v,w >= 0.$$

If a set of vectors in V is orthogonal and each vector is a unit vector, then the set is called **orthonormal**.

Example 3.2. *If S is the set of* 2×2 *symmetric matrices and for all* $U, V \in S$,

$$< U,V >= u_{11}v_{11} + u_{12}v_{12} + u_{21}v_{21} + u_{22}v_{22},$$

$$\left\{ \begin{pmatrix} 0 & 1 \\ 0 & 0 \end{pmatrix}, \begin{pmatrix} 1 & 0 \\ 0 & 1 \end{pmatrix}, \begin{pmatrix} 0 & 0 \\ 1 & 0 \end{pmatrix} \right\},$$

is an orthogonal set in S but is not an orthonormal set in S. The set

$$\left\{ \begin{pmatrix} 0 & 1 \\ 0 & 0 \end{pmatrix}, \begin{pmatrix} \frac{1}{\sqrt{2}} & 0 \\ 0 & \frac{1}{\sqrt{2}} \end{pmatrix}, \begin{pmatrix} 0 & 0 \\ 1 & 0 \end{pmatrix} \right\},$$

is an orthonormal set in S.

3.2   Singular Values

When studying data in which each data point has a large number of attributes, it may be helpful to be able to reduce or compress the number of attributes.

That is, we may start with data points in $\mathbb{R}^n$ and want to project the data points into $\mathbb{R}^m$ where $n > m$.

In order to do so, it is important to reflect on which of the n attributes are significant to the data set. We begin this investigation with a review of some key linear algebra terms, linear independence, span, and basis, and an interpretation of the visual aspects of eigenvectors.

Recall that a set of vectors $\{v_1, v_2, \ldots, v_m\}$ is **linearly independent** if and only if the system of equations

$$k_1 v_1 + k_2 v_2 + \cdots + k_m v_m = \vec{0}$$

has only the trivial solution, $k_1 = k_2 = \cdots = k_m = 0$. Otherwise, we say that the set is **linearly dependent**. If the vectors $v_1, v_2, \ldots, v_m \in \mathbb{R}^n$, then this system can be written as

$$Ax = b$$

$$\begin{pmatrix} v_1 & v_2 & \cdots & v_m \end{pmatrix} \begin{pmatrix} k_1 \\ k_2 \\ \vdots \\ k_m \end{pmatrix} = \begin{pmatrix} 0 \\ 0 \\ \vdots \\ 0 \end{pmatrix}.$$

This is equivalent to saying that the matrix A, whose columns are $v_1, v_2, \ldots, v_m$, is invertible.

If w is a vector in $\mathbb{R}^n$ and there exists constant $k_1, k_2, \ldots, k_m$ such

$$w = k_1 v_1 + k_2 v_2 + \cdots + k_m v_m,$$

we say that w can be written as a **linear combination** and is in the **span** of the vectors $v_1, v_2, \ldots, v_m$.

The set $S = \{v_1, v_2, \ldots, v_m\}$ is a **basis** for a vector space V if S is linearly dependent and if every vector in V is in the span of S.

Example 3.3. *The set $\{(1,0,0), (0,1,0), (0,0,1)\}$ is called the **standard basis** for $\mathbb{R}^3$.*

In fact, this basis is an orthonormal basis for $\mathbb{R}^3$. However, any set of three vectors that are linearly independent can serve as a basis for $\mathbb{R}^3$.

For example, since the matrix

$$A = \begin{pmatrix} 1 & 4 & 7 \\ 2 & 5 & 8 \\ 3 & 6 & 1 \end{pmatrix}$$

is invertible, the system $Ax = 0$ has only the trivial solution and the system $Ax = b$ has a solution for all $b \in \mathbb{R}^3$.

Thus the set of vectors $\{(1,2,3), (4,5,6), (7,8,1)\}$, is a basis for $\mathbb{R}^3$, is linearly independent and spans $\mathbb{R}^3$. Notice that although this set is a basis for $\mathbb{R}^3$, it is neither orthonormal nor orthogonal.

Let A be an $n \times n$ matrix and $T_A : \mathbb{R}^n \to \mathbb{R}^n$ be a linear transformation with standard matrix A, defined as $T_A(x) = Ax$. Then, based on the definition of an eigenvalue λ

$$Ax = \lambda x,$$

any basis vector x for the eigenspace corresponding to the eigenvector λ, also called an eigenvector corresponding λ, will be dilated or contracted by a factor of λ under the transform T_A; however, the direction of x will remain unchanged.

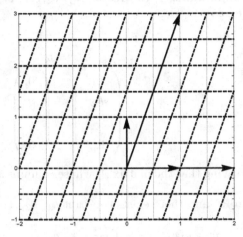

FIGURE 3.1
Example of affects of a linear transformation.

Example 3.4. *Consider the transformation $T_A : \mathbb{R}^2 \to \mathbb{R}^2$, such that*

$$T_A(x) = Ax, \text{ where } A = \begin{pmatrix} 2 & 1 \\ 0 & 3 \end{pmatrix},$$

Under the transformation $T_A(x)$, the standard basis $\{(1,0), (0,1)\}$ is changed to $\{(1,3), (2,0)\}$, as seen in Figure 3.1.

Recall that the **span** of a vector in $\mathbb{R}^2$ is any vector that lies on the same line as the original vector. Most vectors will not remain on their own span under a linear transformation such as T_A.

An example of a vector, it's span, and the vector under $T_A(x)$ can be seen in Figure 3.2. Note that the eigenvectors of A, $x_1 = \{1,1\}$ and $x_2 = \{1,0\}$ will remain on in their span under $T_A(x)$.

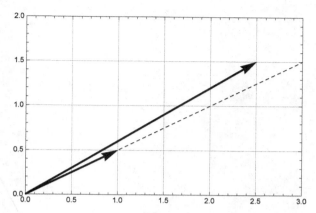

FIGURE 3.2
An example of a vector, it's span, and the vector under $T_A(x)$.

We have seen the importance of eigenvalues and eigenvectors in application in Chapters 1 and 2. If a matrix is not square, no such values can be determined. In this chapter, we will introduce **singular values**, the analogous values to eigenvalues for non-square matrices, and their importance in data science.

Notice in Figure 3.1 that the grid lines under transformation $T_A(x)$ are no longer orthogonal. If the transformed grid lines remain orthogonal then we call this transformed grid an **orthogonal grid**.

Example 3.5. *Consider the transformation $T_A : \mathbb{R}^2 \to \mathbb{R}^2$, such that*

$$T_A(x) = Ax, \text{ where } A = \begin{pmatrix} 2 & 1 \\ 0 & 3 \end{pmatrix},$$

the same transformation and matrix as in Example 3.2. Notice from Figure 3.3 that the standard basis vector, v_1 and v_2, are not orthogonal under $T_A(x)$.

An angle of $\theta \approx 1.28$ radians can be determined such that when v_1 and v_2 are rotated by θ and then the transformation $T_A(x)$ is applied, the vectors

$$u_1 = A \begin{pmatrix} \cos(\theta) & -\sin(\theta) \\ \sin(\theta) & \cos(\theta) \end{pmatrix} v_1 \text{ and } u_2 = A \begin{pmatrix} \cos(\theta) & -\sin(\theta) \\ \sin(\theta) & \cos(\theta) \end{pmatrix} v_2$$

are orthogonal. The norm of the resulting vectors, u_1 and u_2,

$$\sqrt{7 + \sqrt{13}} \text{ and } \sqrt{7 - \sqrt{13}},$$

are the singular values of A. The singular values also corresponding to the square root of the eigenvalues of AA^T.

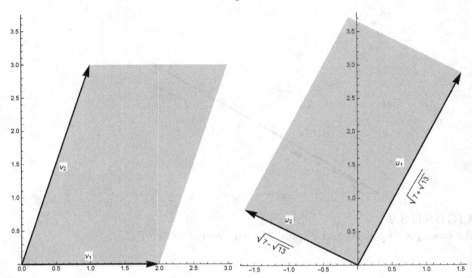

FIGURE 3.3
Example of vectors under a linear transformation that create orthogonal vectors.

3.3   Singular Value Decomposition

The term singular value relates to the distance of the given matrix to a singular matrix. We will explore the concept of singular values in the context of the **Singular Value Decomposition (SVD)** of a matrix.

The **orthogonal components** of a vector v in a n−dimensional vector space W are v_w and $v_{w\perp}$ where $v_w \in W$ and $v_{w\perp} \in W^\perp$ and $v = v_w + v_{w\perp}$.

Given an orthonormal basis $\{v_1, v_2, \cdots, v_n\}$ for W,

$$v_w = Proj_W(v) = <v_1, v> v_1 + <v_2, v> v_2 + \cdots + <v_n, v> v_n.$$

Using the theory above, and the visualization from Example 3.2, for any vector x in $\mathbb{R}^2$, $x = (v_1 \cdot x)v_1 + (v_2 \cdot x)v_2$, where v_1 and v_2 are the standard basis vectors for $\mathbb{R}^2$. Thus,

$$Ax = A(v_1 \cdot x)v_1 + A(v_2 \cdot x)v_2 = u_1\sigma_1(v_1 \cdot x) + u_2\sigma_2(v_2 \cdot x),$$

where σ_i, $1 \leq i \leq n$ are the singular values of A.

Noting that for any two vectors u and w in $\mathbb{R}^2$, $u \cdot w = u^T w$, we can say that

$$Ax = u_1\sigma_1(v_1^T x) + u_2\sigma_2(v_2^T x).$$

More generally, a $m \times n$ matrix A, can be decomposed into the product

$$A = U\Sigma V^T$$

where U is a $m \times m$ orthogonal matrix whose columns form an orthonormal basis for $\mathbb{R}^m$ and V is a $n \times n$ orthogonal matrix whose columns form an orthonormal basis for $\mathbb{R}^n$ and $\Sigma_{ii} = \sigma_i$ and $\sigma_{ij} = 0$ otherwise.

Since the singular values of a matrix, A, are related to the eigenvalues of the symmetric matrix $A^T A$, the Gramian matrix, it is important to note the following property.

Theorem 10. *For any $m \times n$ matrix A, the eigenvalues of the symmetric matrices $A^T A$ and AA^T are all non-negative real numbers.*

Proof. If $A^T A$ has eigenvalue λ with corresponding eigenvector x then $A^T Ax = \lambda x$. Then $x^T A^T Ax = \lambda x^T x$ and $||Ax||^2 = \lambda ||x||^2$. Therefore,

$$\lambda = \frac{||Ax||^2}{||x||^2} \geq 0.$$

A similar argument can be made for the symmetric matrix $A^T A$. $\qquad\square$

Example 3.6. *Let $A = \begin{pmatrix} 2 & 1 & 0 \\ 1 & 2 & 0 \end{pmatrix}$, then*

$$S = A^T A = \begin{pmatrix} 5 & 4 & 0 \\ 4 & 5 & 0 \\ 0 & 0 & 0 \end{pmatrix}.$$

The eigenvalues, σ^2, of S are 9, 1, and 0, so the positive singular values of A are $\sigma_1 = 3$ and $\sigma_2 = 1$.

The columns of matrix V formed by eigenvectors of S,

$$v_1 = \begin{pmatrix} \frac{1}{\sqrt{2}} \\ \frac{1}{\sqrt{2}} \\ 0 \end{pmatrix}, v_2 = \begin{pmatrix} -\frac{1}{\sqrt{2}} \\ \frac{1}{\sqrt{2}} \\ 0 \end{pmatrix}, \text{ and } v_3 = \begin{pmatrix} 0 \\ 0 \\ 1 \end{pmatrix}$$

form an orthonormal basis for $\mathbb{R}^3$.

Since $AV = U\Sigma$. We can determine the columns of U by computing

$$u_1 = \frac{Av_1}{\sigma_1} = \begin{pmatrix} \frac{1}{\sqrt{2}} \\ \frac{1}{\sqrt{2}} \end{pmatrix} \text{ and } u_2 = \frac{Av_2}{\sigma_2} = \begin{pmatrix} -\frac{1}{\sqrt{2}} \\ \frac{1}{\sqrt{2}} \end{pmatrix},$$

which form an orthonormal basis for $\mathbb{R}^2$. The singular value decomposition of

$$A = U\Sigma V^T,$$

$$A = \begin{pmatrix} \frac{1}{\sqrt{2}} & -\frac{1}{\sqrt{2}} \\ \frac{1}{\sqrt{2}} & \frac{1}{\sqrt{2}} \end{pmatrix} \begin{pmatrix} 9 & 0 & 0 \\ 0 & 1 & 0 \end{pmatrix} \begin{pmatrix} \frac{1}{\sqrt{2}} & \frac{1}{\sqrt{2}} & 0 \\ -\frac{1}{\sqrt{2}} & \frac{1}{\sqrt{2}} & 0 \\ 0 & 0 & 1 \end{pmatrix}.$$

Recall from Chapter 2 that all symmetric matrices are orthogonally diagonalizable. Thus for any $m \times n$ matrices A, $S = A^T A$ is a $n \times n$ symmetric matrix and we can find an orthogonal matrix P such that

$$S = PDP^T.$$

This shows an interesting alignment between eigenvalues of S and singular values of A. Assume that the eigenvalues of S are $\lambda_1 \geq \lambda_2 \geq \cdots \geq \lambda_n$ and the corresponding singular values of A are $\sigma_1 \geq \sigma_2 \geq \cdots \geq \sigma_n$. Then,

$$S = PDP^T = \sum_{i=1}^{n} \lambda_i v_i v_i^T$$

$$A = U\Sigma V^T = \sum_{i=1}^{n} \sigma_i u_i v_i^T.$$

3.4 Compression of Data Using Principal Component Analysis (PCA)

In data science, you may wish to compare m records with n attributes in a variety of ways. There are many clustering techniques, but in all of these techniques the data scientist might also be interested in visualizing the data. In order to do this, one should consider how to transform their data from $\mathbb{R}^n$ to $\mathbb{R}^2$ or $\mathbb{R}^3$.

One way to think about this is that you are projecting your data from $\mathbb{R}^n$ into another space, such as $\mathbb{R}^2$ or $\mathbb{R}^3$.

We can do this using SVD by limiting the number of singular values used to reconstruct the original matrix.

Recall that if A is a $m \times n$ matrix then the singular value decomposition of $A = U\Sigma V^T$ with singular values $\sigma_1, \sigma_2 \ldots, \sigma_n$, Σ is a $m \times n$ matrix such that $\Sigma_{ii} = \sigma_i$.

If a data scientist wishes to transform the data from $\mathbb{R}^n$ to $\mathbb{R}^q$, then they could determine the q most dominant eigenvalues and define a $q \times n$ matrix $\tilde{\Sigma}$ such that $\tilde{\Sigma}_{ii} = \sigma_i$.

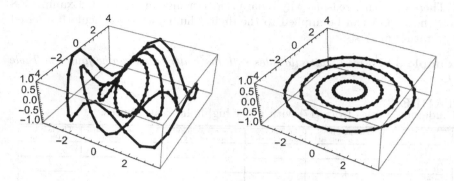

FIGURE 3.4
Example of a projection from $\mathbb{R}^3$ into $\mathbb{R}^2$.

In turn, matrix $\tilde{U}$ would then become a $m \times q$ matrix, where $\tilde{U}_{i,j} = U_{i,j}$ for $1 \le i \le m, 1 \le j \le q$.

The rows of $\tilde{U}$ are vectors in $\mathbb{R}^q$ and are the focus of investigation in Principal Component Analysis (PCA).

Example 3.7. *Let* $A = \begin{pmatrix} 1 & 2 & 1 & 0 & 0 \\ 0 & 0 & 1 & 2 & 1 \\ 0 & 1 & 0 & 0 & 2 \end{pmatrix}$. *Then*

$$S = A^T A = \begin{pmatrix} 1 & 2 & 1 & 0 & 0 \\ 2 & 5 & 2 & 0 & 2 \\ 1 & 2 & 2 & 2 & 1 \\ 0 & 0 & 2 & 4 & 2 \\ 0 & 2 & 1 & 2 & 5 \end{pmatrix}$$

has eigenvalues $\lambda_1 = 9$, $\lambda_2 = 5$, $\lambda_3 = 3$, *and* $\lambda_4 = \lambda_5 = 0$. *The non-zero eigenvalues correspond to the singular values of* A *are* $3, \sqrt{5},$ *and* $\sqrt{3}$.

If PCA is applied to limit our focus to the two largest singular values, 3 and $\sqrt{5}$, *then the subsequently columns of* $\tilde{U}$ *are the corresponding set of orthonormal eigenvectors.*

$$\tilde{U} = \begin{pmatrix} \frac{1}{\sqrt{3}} & -\frac{1}{\sqrt{2}} \\ \frac{1}{\sqrt{3}} & \frac{1}{\sqrt{2}} \\ \frac{1}{\sqrt{3}} & 0 \end{pmatrix}.$$

There are many reasons why a data scientist may apply PCA. Example 3.8 shows how PCA can be applied to the digital humanities, particularly related to literature analysis.

Example 3.8. *Five Cinderella tales with their attributes can be found in Table 3.1.*

TABLE 3.1
Cinderella stories from [17] with a few highlighted attributes.

	The heroine's name means ash	The heroine picks lentils	Magical animals are present	Action takes place in a church	The heroine buries bones
Zezzola (Italy)	1	0	1	0	0
Aschenputtel (Germany)	1	1	1	0	1
Rashin-Coatie (Scotland)	0	0	1	1	0
Ye Xian (China)	0	0	0	0	1

We construct an adjacency matrix, A, with rows representing each tale and columns representing the traits present in each tale.

$$A = \begin{pmatrix} Zezzola & 1 & 0 & 1 & 0 & 0 \\ Ashenputtel & 1 & 1 & 1 & 0 & 1 \\ Rahin - Coatie & 0 & 0 & 1 & 1 & 0 \\ YeXian & 0 & 0 & 0 & 0 & 1 \end{pmatrix}.$$

Since we are interested in focusing on the relationship between the tales, and not the traits, we construct,

$$S = AA^T = \begin{pmatrix} 2 & 2 & 1 & 0 \\ 2 & 4 & 1 & 1 \\ 1 & 1 & 2 & 0 \\ 0 & 1 & 0 & 1 \end{pmatrix}$$

which has eigenvalues equal to $\{5.8737, 1.85026, 1, 0.276044\}$.

For this example, let's assume that we wish to project our visualization of these tales into $\mathbb{R}^2$ *and thus we wish to limit the eigenvalues of S to the two largest eigenvalues, principal components,* $\lambda_1 = 5.8737$ *and* $\lambda_2 = 1.85026$. $\tilde{A} = \tilde{U}\tilde{\Sigma}\tilde{V}^T$ *with*

$$\tilde{\Sigma} = \begin{pmatrix} \sqrt{5.8737} & 0. \\ 0. & \sqrt{1.85026} \end{pmatrix}$$

$$\tilde{V}^T = \begin{pmatrix} 0.528603 & 0.325381 & 0.665063 & 0.136459 & 0.392144 \\ 0.0818573 & 0.288821 & -0.464793 & -0.546651 & 0.628508 \end{pmatrix}$$

and

$$\tilde{U} = \begin{pmatrix} 0.492523 & -0.281521 \\ 0.788584 & 0.392866 \\ 0.330719 & -0.743577 \\ 0.161804 & 0.462056 \end{pmatrix}.$$

Notice that the columns of $\tilde{V}$ corresponding to the eigenvectors associated with λ_1 and λ_2 and the ith column of $\tilde{U}$ is the ith column of U.

It is also important to notice that $\tilde{U}$ is a $m \times 2$ matrix and thus for each of the $m = 5$ tales there is associated ordered pair in $\tilde{U}$.

Figure 3.5 shows a visualization of the Cinderella tales using $\tilde{U}$.

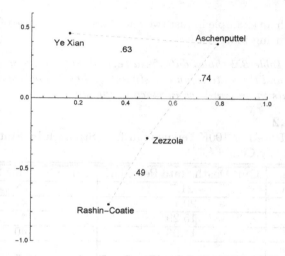

FIGURE 3.5

Coordinates from $\tilde{U}$ representing the Tales from Table 3.1 in $\mathbb{R}^2$.

3.5 PCA, Covariance, and Correlation

When studying large data sets with many variables, we may wish to determine patterns in the data and if there are variables that have a strong correlation with one another. In this section, we will look at how PCA can be used to create a **covariance matrix** and **correlation matrix**.

The covariance matrix of a data set A, $Cov(A)$, captures the variance and linear correlation in multivariate data. Covariance is a measure of how much

m records of data, each with N attributes or pieces of information, change together.

Note that $Cov(A)_{i,i}$ represents the variance, σ_i^2, in the ith data record.

$$
Cov(A) = \begin{pmatrix}
\sum \frac{x_1^2}{N} & \sum \frac{x_1 x_2}{N} & \sum \frac{x_1 x_3}{N} & \cdots & \sum \frac{x_1 x_m}{N} \\
\sum \frac{x_2 x_1}{N} & \sum \frac{x_2^2}{N} & \sum \frac{x_2 x_3}{N} & \cdots & \sum \frac{x_2 x_m}{N} \\
\sum \frac{x_3 x_1}{N} & \sum \frac{x_3 x_2}{N} & \sum \frac{x_3^2}{N} & \cdots & \sum \frac{x_3 x_m}{N} \\
\vdots & \vdots & \vdots & \ddots & \vdots \\
\sum \frac{x_m x_1}{N} & \sum \frac{x_m x_2}{N} & \sum \frac{x_m x_3}{N} & \cdots & \sum \frac{x_m^2}{N}
\end{pmatrix}.
$$

We begin with an example in just two variables to get a sense of how variance can be modeled using matrices.

Example 3.9. *Table 3.2 shows data from several states on gun deaths per 100 thousand people and the state's gun law strength, where a lower gun law strength number represents more gun laws in place in the state.*

TABLE 3.2
Gun Death Rate Per 100K versus Gun Law Strength by State (Violence Policy Center [35]).

State	Gun Death Rate Per 100K	2019 Gun Law Strength
Alabama	21.7	38
Alaska	20.74	42
Arizona	15.29	45
Arkansas	18.96	40
California	7.45	1
Colorado	15.14	14
Connecticut	4.91	3
Delaware	11.55	11
Florida	12.81	22
Georgia	15.72	32
Hawaii	4.03	5
Idaho	16.61	48
Illinois	10.78	8
Indiana	14.71	28
Iowa	8.62	19
Kansas	14.65	43
Kentucky	16.81	46
Louisiana	21.31	32
Maine	10.32	34

There are several pre-processing techniques that can be done to the data to allow the data to be roughly symmetric across the origin. One of these techniques *is called **zero-centering**, where the data is translated by the mean so that the mean is at the origin.*

*Another technique is to **standardize** the data so that the data has a mean of 0 and a variance of 1.*

Many algorithms tend to do better with symmetric data, particularly when studying the rate of change of the data, and zero-centering does not alter relationships among samples and among components of the same sample.

In this example, the data presented in Table 3.2 is zero-centered before it is analyzed.

Thus, if the rows of A consist of the original vectors { Gun law strength, Death rate per 100K}, the transformation matrix, under zero-centering, B is defined as

$$B_{i,1} = A_{i,1} - 26.8947 \text{ and } B_{i,2} = A_{i,2} - 13.7953 \text{ for } 1 \leq i \leq 19.$$

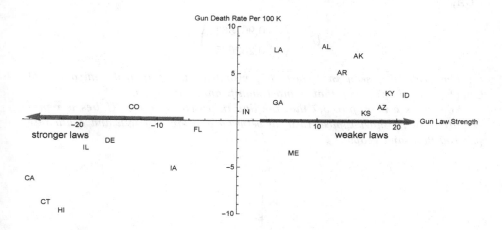

FIGURE 3.6
Gun law strength versus Gun Death Rate Per 100K.

Visually inspecting Figure 3.6, you may notice that there appears to be a positive linear relationship between gun law strength and the gun death rate per 100 thousand people. That is, as the gun law strength gets weaker (higher number), the gun death rate increases in the state.

Inspecting the covariance matrix in this example,

$$Cov(B) = \begin{pmatrix} 255.988 & 64.1189 \\ 64.1189 & 27.1512 \end{pmatrix}.$$

$Cov(B)_{1,1}$ *represents the sample variance in gun death rate per 100K and* $Cov(B)_{2,2}$ *represents the sample variance in the gun law strength.*

$Cov(B)_{1,2} = Cov(B)_{2,1} = 64.1189 > 0$ *indicates that there is a positive correlation between these two variables.*

Principal Component Analysis can be used to decompose the covariance matrix, limiting the singular values to the ones that are most significant.

Example 3.10. *Investigated the covariance matrix from Example 3.9 further,*

$$Cov = \begin{pmatrix} 255.988 & 64.1189 \\ 64.1189 & 27.1512 \end{pmatrix},$$

has two singular values, $\sigma_1 = 272.729$ and $\sigma_2 = 10.4101$. Noting that σ_2 is significantly smaller than σ_1, one might focus on σ_1 to do further exploration into the relationship between gun death rate per 100 thousand people and state gun law strength.

Reconstructing U with only σ_1 taken into consideration, and thus using PCA,

$$\tilde{U} = \begin{pmatrix} -0.967564 \\ -0.252625 \end{pmatrix}.$$

The vector $\tilde{U}$, seen in Figure 3.7, visualizes the principal component of $Cov(B)$ and shows a ray that is most significant in the data.

You can see in Figure 3.7 that the line in which the vector $\tilde{U}$ lies is representative of the most typical ratio of between gun law strength and death rate per 100 thousand people.

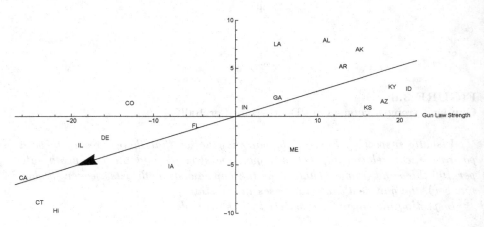

FIGURE 3.7
Gun law strength versus Gun Death Rate Per 100K with the covariance direction represented.

The **correlation matrix** associated with a data set, A, $Cor(A)$, is just a scaling of the covariance matrix,

$$Cor(A)_{i,j} = \frac{Cov(A)_{i,j}}{Cov(A)_{i,i}Cov(A)_{j,j}}$$

for $i \neq j$ and $Cor(A)_{i,j} = 1$ if $i = j$.

It is a good idea to consider using the correlation matrix to study your data, rather than the covariance matrix, if your variables are in scales of very different magnitudes.

These ideas can be expanded to multiple regression techniques using Principal Component Analysis, more commonly called **Principal Component Regression**.

Unlike Example 3.9, where there was only one dependent variable, x, and one independent variable, y, and $y = ax + b + e$, with e representing the errors in the residuals, it is possible to use principal component regression in a more complex model with k independent variables. In this case, the model becomes

$$y = Xb$$

where y is an $n \times 1$ vector, X is an $n \times k$ matrix.

Like Example 3.9, one should consider zero-centering or standardizing the data before finding the covariance and correlation matrices. The eigenvectors of the covariance matrix, typically stored in the matrix U in the singular value decomposition, give a means of understanding the direction of the data based on that variable. While the singular values, eigenvalues, of the covariance matrix give a sense of how much the data varies in that direction.

Because of the significance of the singular values in determining variance in the data, we must determine how many principal components, p, to include in the model. The best way to do this is by inspecting the correlation matrix and omitting small singular values of the correlation matrix, which typically have values significantly less than 1.

In determining a least square regression model with all k factors integrated into the model, we wish to determine b, where

$$b = (X^T X)^{-1} X^T y.$$

In order to follow this procedure, X must be an orthogonal matrix, and thus $X^T X$ is diagonal.

Unfortunately, most datasets are correlated and thus X is not an orthogonal matrix. This leads most data scientists to proceed with some sort of variable selection and thus replacing the k correlated variables of X with p uncorrelated variables, represented by an $n \times p$ matrix $\tilde{X}$.

The Principal Component Regression model $\hat{y} = \tilde{X}b$ can be determined by reducing the principal components used to $p < k$ such that $\tilde{X} = \tilde{U}$ and

$$b = (\tilde{X}^T \tilde{X})^{-1} \tilde{X}^T y.$$

TABLE 3.3

2020 Population and Number of Registered Guns by State (World Population Review [42]).

State	Population in 2020	Number of Registered Guns
Alabama	4,908,621	161,641
Alaska	734,002	15,824
Arizona	7,378,494	179,738
Arkansas	3,038,999	79,841
California	39,937,489	344,642
Colorado	5,845,526	92,435
Connecticut	3,563,077	82,400
Delaware	982,895	4,852
Florida	21,992,985	343,288
Georgia	10,736,059	190,050
Hawaii	1,412,687	7,859
Idaho	1,826,156	49,566
Illinois	12,659,682	146,487
Indiana	6,745,354	114,019
Iowa	3,179,849	28,494
Kansas	2,910,357	52,634
Kentucky	4,499,692	81,058
Louisiana	4,645,184	116,831
Maine	1,345,790	15,371

Example 3.11. *In Example 3.9, we used state gun law strength to predict the gun death rate per 100 thousand people in the state. In this example, we will determine a model that will explore how state gun law strength, from Table 3.2, the population of the state, and the number of registered guns in the state can be used to predict the gun death rate per 100 thousand people. The population in 2020 and number of registered guns by state can be found in Table 3.3.*

In Example 3.9, the data was zero-centered; however, when the population and number of guns registered by state are also included, we begin to have variables of very different magnitudes. In this case, we will standardize the data before we begin.

After standardizing the data, the corresponding correlation matrix will be

$$
Cor = \begin{pmatrix}
 & DeathRate & GunLawStrength & Population & \#ofGuns \\
DeathRate & 1 & 0.769 & -0.25 & -0.002 \\
GunLawStrength & 0.769 & 1 & -0.408 & -0.188 \\
Population & -0.25 & -0.408 & 1 & 0.891 \\
\#ofGuns & -0.002 & -0.188 & 0.891 & 1
\end{pmatrix}
$$

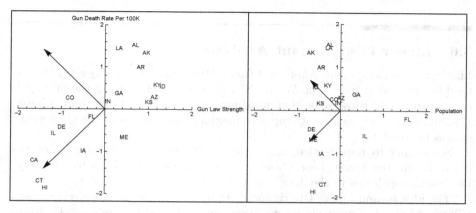

FIGURE 3.8
Dependent Variable Versus Respective Independent Variables in the Model.

with singular values $\sigma_1 = 2.27749, \sigma_2 = 1.43598, \sigma_3 = 0.212222$, and $\sigma_4 = 0.0743121$. We proceed with the regression using only the two largest principal components.

Let X_1 be the state gun law strength (with mean $\bar{x}_1$ and standard deviation s_1), X_2 represents the population by state (with mean $\bar{x}_2$ and standard deviation s_2), X_3 represents number of registered guns by state (with mean $\bar{x}_3$ and standard deviation s_3), and y the number of gun deaths by state (with mean $\bar{x}_y$ and standard deviation s_y).

We standardize each factor in order to get each variable on a similar scale and create X, the independent variable data matrix with columns consisting of standardized X_1, X_2, X_3.

When PCA is performed on X, similar to Example 3.9, with only 2 singular values, $b = \{-2.68968, 2.99994\}$. Thus

$$\frac{\widehat{gun\ deaths} - \bar{x}_y}{s_y} = -2.68968 \left(\frac{law\ strength - \bar{x}_1}{s_1} \right)$$
$$- 2.99994 \left(\frac{population - \bar{x}_2}{s_2} \right).$$

Figure 3.8 shows an additional visualization of gun deaths versus gun law strength and state population size with normalized data. The vectors on these graphs are the eigenvectors associate with the represented principal components.

3.6   Linear Discriminant Analysis

Like Principal Component Analysis, **Linear Discriminant Analysis** (LDA) is used for dimension reduction. While PCA focuses on determining the principal component axes, those axes that maximize the variance of the data, Linear Discriminant Analysis also attempts to maximize the spread of the clusters or classes in the data.

Something to note is that many times with PCA the data may not be originally clustered into classes; however, it is necessary with LCA to originally be identifying classes in the data.

The idea behind Linear Discriminant Analysis is to choose a projection that will spread out the known classes while minimizing the scatter, or spread, within each class.

Imagine that you have two classes of data in $\mathbb{R}^2$ and you wish to project the data onto a line. If the classes are indeed unique, the choice of line on which you choose to project does make a difference in terms of between class spread.

For example, in Figure 3.9, the projection of the two classes of data, represented by circles and squares, onto the x-axis and the line $y = 2x$ do not do a great job of spreading out the two classes, whereas a projection onto the line $y = -x$, seen in Figure 3.10, completely separates the two classes.

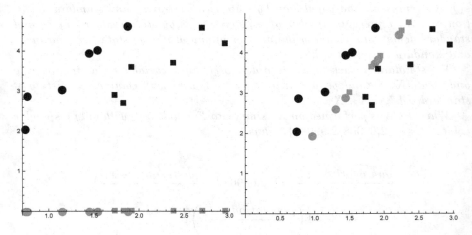

FIGURE 3.9
Example of projection onto the x-axis and the line $y = 2x$.

Let's assume that there are n classes, $C_1, C_2, \ldots, C_n$, where the ith class is of size N_i. In addition, assume that each vector, x, representing an object, has m entries and that the line of projection will be represented by the unit vector v.

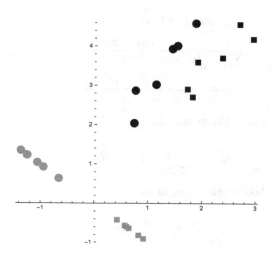

FIGURE 3.10
Example of projection onto the line $y = -x$.

Then the

$$\text{projection of } x \text{ onto } v = \frac{<v, x> v}{||v||^2}$$

and the distance that x is from the origin is

$$v^T x.$$

Define $\tilde{\mu}_i$ as the mean of the projections of vectors in classes C_i onto the vector v. Thus

$$\tilde{\mu}_i = \sum_{x \in C_i} \frac{v^T x}{N_i} = v^T \sum_{x \in C_i} \frac{x}{N_i} = v^T \mu_i,$$

where μ_i is the mean of the vectors in class C_i.

Given two classes C_1 and C_2, we wish to choose a projection that maximizes

$$|\tilde{\mu}_1 - \tilde{\mu}_2|$$

in order to maximize the spread of the classes; however, this does not minimize the spread within classes. The objective of Linear Discriminant Analysis is to find a vector v that takes each of these criteria into account. In order to do this, we will maximize the objective function

$$J(v) = \frac{(\tilde{\mu}_1 - \tilde{\mu}_2)^2}{\sum_{x \in C_1}(v^T x - \tilde{\mu}_1)^2 + \sum_{x \in C_2}(v^T x - \tilde{\mu}_2)^2}$$

$$= \frac{(v^T \mu_1 - v^T \mu_2)^2}{\sum_{x \in C_1}(v^T x - \tilde{\mu}_1)^2 + \sum_{x \in C_2}(v^T x - \tilde{\mu}_2)^2}.$$

Define the within class scatter matrix,

$$S_W = \sum_{i=1}^{n} \sum_{x \in C_i} (x - \mu_i)(x - \mu_i)^T$$

and the between class scatter matrix as

$$S_B = (\mu_1 - \mu_2)(\mu_1 - \mu_2)^T.$$

Then,

$$J(v) = \frac{v^T S_B v}{v^T S_W v}.$$

In order to maximize $J(v)$, we determine where $\frac{\partial J}{\partial v} = 0$.

$$\frac{\partial J}{\partial v} = \frac{(2S_B v)(v^T S_W v) - (2S_w v)(v^T S_B v)}{(v^T S_W v)^2} = 0$$

when $(S_B v)(v^T S_W v) - (S_W v)(v^T S_B v) = 0$.
That is, we wish to find v such that

$$S_B v - \frac{(S_W v)(v^T S_B v)}{v^T S_W v} = 0.$$

If S_W is invertible and $\lambda = \frac{(v^T S_B v)}{v^T S_W v}$, $S_W^{-1} S_B v = \lambda v$ and v is the eigenvector associated with matrix $S_W^{-1} S_B$ and eigenvalue λ.

Another way to think about this, due to the definition of S_B, is that for any vector v, $S_B v$ points in the same direction as $\mu_1 - \mu_2$. Therefore, we wish to determine v such that

$$v = S_W^{-1}(\mu_1 - \mu_2).$$

Example 3.12. *For the data shown in Figure 3.9,* $\mu_1 = \{1.25584, 3.42925\}$ *and* $\mu_2 = \{2.25309, 3.5972\}$

$$S_W = \begin{pmatrix} 2.29589 & 3.62306 \\ 3.62306 & 6.93158 \end{pmatrix} \text{ and } S_B = \begin{pmatrix} 0.994501 & 0.167485 \\ 0.167485 & 0.0282064 \end{pmatrix}.$$

The best line of projection is defined by vector

$$v = S_W^{-1}(\mu_1 - \mu_2) = \{-2.26148, 1.15782\}.$$

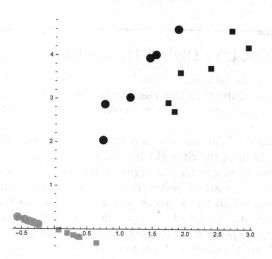

FIGURE 3.11

Data from Figure 3.9 projected onto the vector $v = \{-2.26148, 1.15782\}$.

The concepts described in this section can be generalized for a dataset with n classes.

1. Find each class C_i, $2 \leq i \leq n$, find the mean vector,

$$\mu_i = \sum_{v \in C_i} \frac{v}{N_i}.$$

2. Construct the within class scatter vector S_W, where the entries of S_W are

$$\sum_{i=1}^{n} \sum_{x \in C_i} (x - \mu_i)(x - \mu_i)^T.$$

3. Construct the between-class scatter matrix S_B, where the entries of S_B are

$$\sum_{i=1}^{n} N_i(\mu_i - \mu),$$

where μ is the mean of all of the samples.

4. Determine the eigenvector v of $S_W^{-1} S_B$.

3.7　CASE STUDY: Digital Humanities

Some digital humanists focus on semantics and words that are regularly used or used in a particular context. In this case study, we will focus on characteristics of different versions of a story that are present or absent in order to determine the story's origin.

The Star Husband Tale was a widely known tale among Native American tribes. The basic form of the Star Husband Tale [45] tells of two girls who are sleeping out in the open during the night. While outside, they see two stars, and each girl makes a wish to be married to a star. When they awake, both have been transported to the heavens and are married to the stars as they wished. One of the star husbands is a young man, and the other is an older man. The girls at one point start digging in the heavens and make a hole, and they eventually lower themselves down to earth using a rope.

Dundes [9] discusses narrative elements of 86 versions of the Star Husband Tale coming from 44 tribes, grouped into nine geographical zones: Eskimo, Mackenzie, North Pacific, California, Plateau, Plains, Southeast, Southwest, and Woodlands. In this case study, we will explore the 10 tales in the Woodlands tribe and some of their archetypes and subarchetypes. These archetypes and subarchetypes include how the girl(s) in the folktale travel to the upper world, taboos broken in the upper world, whether the girl(s) in the story dig a hole in the cloud, and how the girl(s) in the story travel home.

Data for a more in-depth study of the Star Husband Tale can be found at Github link 23. Python and R code for this case study can be found at GitHub links 19 and 21 respectively.

The rows of data matrix, M, represent the 10 Woodland tribes and columns represent the archetypes and subarchtypes. If the $(i,j)^{\text{th}}$ entry of M is equal to 1 then archetype j is present in tribe $i's$ tale.

$$
M = \begin{pmatrix}
Ojibawa1 & 1 & 0 & 0 & 0 & 0 & 1 & 0 & 1 & 1 & 0 & 0 & 0 & 1 & 0 & 0 & 1 \\
Ojibawa2 & 0 & 0 & 0 & 0 & 0 & 1 & 0 & 1 & 0 & 0 & 0 & 0 & 1 & 0 & 0 & 1 \\
Ojibawa3 & 1 & 0 & 0 & 0 & 0 & 1 & 0 & 1 & 0 & 1 & 1 & 0 & 1 & 0 & 0 & 1 \\
Ojibawa4 & 1 & 0 & 0 & 0 & 0 & 0 & 0 & 0 & 0 & 0 & 0 & 0 & 0 & 0 & 1 & 1 \\
Ojibawa5 & 1 & 0 & 0 & 0 & 0 & 1 & 0 & 0 & 0 & 1 & 1 & 0 & 1 & 1 & 0 & 0 \\
MicMac1 & 1 & 1 & 1 & 0 & 1 & 0 & 1 & 0 & 0 & 1 & 0 & 1 & 1 & 0 & 0 & 1 \\
MicMac2 & 1 & 1 & 1 & 1 & 1 & 0 & 0 & 0 & 0 & 1 & 0 & 1 & 1 & 0 & 0 & 1 \\
MicMac3 & 1 & 1 & 1 & 0 & 0 & 0 & 0 & 0 & 0 & 0 & 0 & 0 & 0 & 0 & 0 & 0 \\
Passamaquody1 & 1 & 1 & 0 & 1 & 1 & 0 & 1 & 0 & 0 & 1 & 0 & 1 & 1 & 0 & 0 & 1 \\
Passamaquody2 & 1 & 1 & 0 & 1 & 1 & 0 & 0 & 0 & 0 & 1 & 0 & 1 & 1 & 0 & 0 & 1
\end{pmatrix}.
$$

Using only the two largest principal components of M, $\sigma_1 = 6.37802$ and $\sigma_2 = 3.30583$ the matrix

$$\tilde{U} = \begin{pmatrix} -0.24 & -0.47 \\ -0.17 & -0.4 \\ -0.31 & -0.49 \\ -0.13 & -0.08 \\ -0.23 & -0.36 \\ -0.43 & 0.24 \\ -0.44 & 0.26 \\ -0.14 & 0.13 \\ -0.43 & 0.25 \\ -0.41 & 0.2 \end{pmatrix}.$$

If we use the rows of $\tilde{U}$ as the coordinates of each of the tales/tribes, we can visualize the tales in $\mathbb{R}^2$. Figure 3.13 shows a visual representation of the these tales with the tribes labeled.

It is interesting to note that, for the most part, the tales from the five Ojibawa tribes and the tales for the Passamaquody tribes are close to one another in this visualization.

One can also compare the results with the actual physical location of the tribes in Figure 3.12, noting that it would be common for the Passamaquody

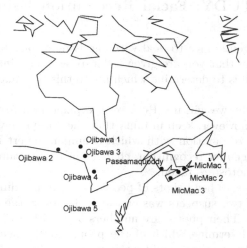

FIGURE 3.12
Location of Woodlands tribes.

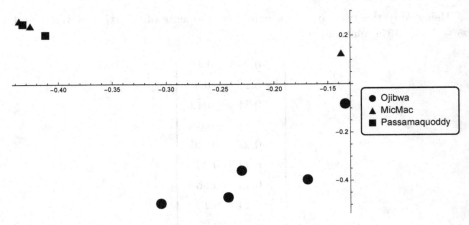

FIGURE 3.13
Visual representation of Woodland tribes using PCA.

tribes to share hunting grounds and thus stories with a nearby tribe such as a MicMac tribe.

3.8 CASE STUDY: Facial Recognition Using PCA and LDA

Law enforcement agents have to deal with the issue of facial recognition on a daily basis. Let's say that you are given a database of criminal pictures and a new photo. Your job is to determine which photo this new photo resembles the most.

In this case study, we will use PCA on the pixel data from each photo to cluster the photos; however, keep in mind that there are many factors that are important in facial recognition with which a human expert may do a better job. These include gender recognition, changes in hair features, and even the position that the photos are taken.

Figure 3.14 shows 10 head-shots of people that are in a line up. In a recent crime, the photo of two suspects was captured; however, the resolution of the photos is very poor. Their photos are numbers 5 and 10 in Figure 3.14.

Your job is to determine which of the people in the head-shots, 1-4 and 6-9, from Figure 3.14 closely resembles the suspects' photos. If you are able to narrow your search down to subcategories, you could look at a smaller subset of photos to find your suspect.

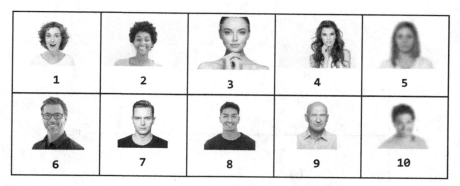

FIGURE 3.14
10 Head-shots for Facial Recognition Study [14].

The nine-byte representation of pixels in the middle of each of the 5 head-shots, 1-4, in Figure 3.14 can be found in the columns of matrix A. We will use a small version of the image data from the center of each photo for this case study, which can be found at GitHub link 24.

$$A = \begin{pmatrix} 97 & 125 & 106 & 172 & 238 \\ 96 & 93 & 100 & 185 & 239 \\ 81 & 72 & 93 & 200 & 240 \\ 87 & 70 & 103 & 171 & 235 \\ 95 & 54 & 97 & 187 & 236 \\ 86 & 55 & 88 & 203 & 237 \\ 48 & 58 & 100 & 169 & 232 \\ 67 & 47 & 96 & 187 & 233 \\ 90 & 51 & 87 & 203 & 234 \end{pmatrix}.$$

Using the two largest principal components of $A^T A$ to construct $\tilde{U}$, we can visualize a representation of these photos in $\mathbb{R}^2$, seen in Figure 3.15, with the entries of the eigenvector associated with the largest principal component on the x-axis and those associated with the second largest principal component on the y-axis.

Unfortunately, if you are unable to narrow the group into subgroups initially, distinctive features may not be as easily detected by PCA, with only nine data pixels per picture. We see that the matrix, B, which includes nine data points for each of the 10 head-shots,

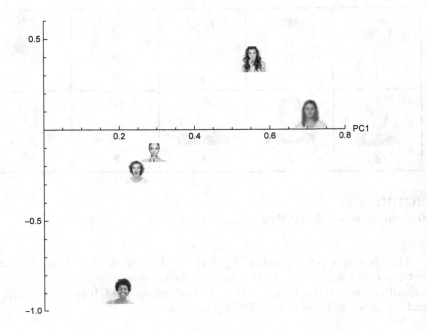

FIGURE 3.15
Visualization of photos 1-5 base on PCA.

$$B = \begin{pmatrix} 97 & 125 & 106 & 172 & 238 & 158 & 216 & 191 & 91 & 193 \\ 96 & 93 & 100 & 185 & 239 & 153 & 218 & 218 & 94 & 193 \\ 81 & 72 & 93 & 200 & 240 & 147 & 218 & 244 & 98 & 193 \\ 87 & 70 & 103 & 171 & 235 & 156 & 215 & 193 & 86 & 193 \\ 95 & 54 & 97 & 187 & 236 & 154 & 212 & 225 & 92 & 192 \\ 86 & 55 & 88 & 203 & 237 & 151 & 209 & 227 & 98 & 192 \\ 48 & 58 & 100 & 169 & 232 & 159 & 207 & 178 & 82 & 192 \\ 67 & 47 & 96 & 187 & 233 & 158 & 201 & 203 & 89 & 192 \\ 90 & 51 & 87 & 203 & 234 & 156 & 195 & 215 & 97 & 191 \end{pmatrix},$$

one can see from Figure 3.16 that the #5 suspect no longer is closest in resemblance to Figures 1-4 in the lineup. However, if there were over 10000 pixels per suspect, as seen in Figure 3.17, more distinction between groups is recognized with PCA. This data can be found at GitHub link 25.

In this analysis, using PCA, you may not be able to make an assumption about which subgroup, 1-4 or 6-8, the two suspects belong to, nor can an as-

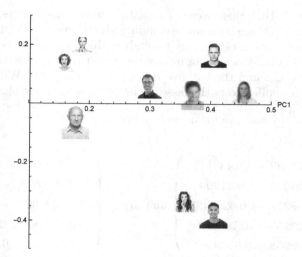

FIGURE 3.16
PCA with only 9 pixels per suspect with data.

FIGURE 3.17
PCA with over 10000 pixels per suspect.

sumption be made that there were two distinct subgroups, classes, among the dataset. However, if there is an understanding about how the 10 photos can be placed into two classes, a better way to analyze the photos may be with LDA.

When applying LDA, we begin by assuming that the first five columns of B fit into one class, C_1, and the last five into a second class, C_2. With the larger data set, PCA is applied to both classes in order to project the classes into $\mathbb{R}^2$.

$$S_w = \begin{pmatrix} 2 & 0 \\ 0 & 2 \end{pmatrix} \text{ and the mean vectors are}$$

$$\tilde{\mu}_{C_1} = \begin{pmatrix} -0.404322 & -0.433172 \\ -0.423223 & -0.419999 \\ -0.468211 & -0.0258423 \\ -0.423255 & 0.795089 \\ -0.508958 & 0.0559352 \end{pmatrix} \text{ and } \tilde{\mu}_{C_2} = \begin{pmatrix} -0.393657 & 0.00717916 \\ -0.494921 & -0.587535 \\ -0.433602 & -0.161582 \\ -0.437483 & 0.792504 \\ -0.469771 & 0.0240795 \end{pmatrix}.$$

Projecting onto the vector $v = \{0.0001464, -0.01026\}$ produces Figure 3.18.

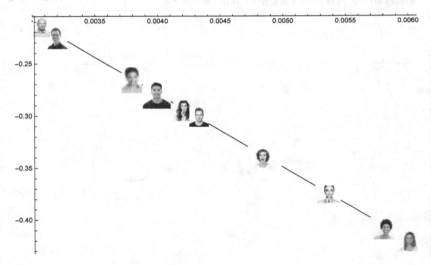

FIGURE 3.18
Projecting classes onto v.

Notice that the two classes are not completely well defined, separated, with this projection; however, there are some clear distinctions between the classes.

Similar ideas can be applied to handwriting samples and other image recognition problems.

Python and R code related to this Case Study can be found at GitHub links 20 and 22 respectively.

3.9 Exercises

1. Define the transformation $T_A : \mathbb{R}^2 \to \mathbb{R}^2$ with standard matrix

$$A = \begin{pmatrix} 1 & 3 \\ 2 & 4 \end{pmatrix}.$$

   a. Graph the standard basis vectors, $v_1 = \{1,0\}$ and $v_2 = \{0,1\}$, transformed under T_A and determine if the resulting vectors orthogonal.

   b. Find an angle of rotation, θ, such u_1 and u_2 are orthogonal, where

$$u_1 = T_A \begin{pmatrix} \cos(\theta) & -\sin(\theta) \\ \sin(\theta) & \cos(\theta) \end{pmatrix} v_1,$$

$$u_2 = T_A \begin{pmatrix} \cos(\theta) & -\sin(\theta) \\ \sin(\theta) & \cos(\theta) \end{pmatrix} v_2.$$

   c. Use u_1 and u_2 to find the singular values of A.

   d. Use AA^T to find the singular values of A.

2. Find the singular values of

$$\begin{pmatrix} 4 & 0 & 3 \\ 0 & 2 & 0 \end{pmatrix}.$$

3. Find the singular values and eigenvalues of

$$\begin{pmatrix} 1 & 2 & 3 \\ 2 & -1 & 0 \\ 3 & 0 & 4 \end{pmatrix}.$$

4. Prove that if A is a square matrix then the eigenvalues of A are equal to the singular values of A.

5. Carl just got a new dog and is curious about its genetic makeup and the genetic relationship between many pure bred dogs. Matrix A shows a small portion of the genetic makeup (specifically haplotypes) of Carl's new dog and that of several pure bred dogs. The haplotype names in order of column (if you are interested are) A2, A11, A16, A17, A18, A19, B1A.

$$A = \begin{pmatrix} 0 & 1 & 0 & 0 & 1 & 1 & 1 \\ 0 & 1 & 0 & 0 & 1 & 0 & 0 \\ 0 & 0 & 0 & 1 & 0 & 1 & 1 \\ 1 & 1 & 0 & 0 & 0 & 0 & 1 \\ 0 & 1 & 1 & 1 & 0 & 1 & 1 \\ 0 & 0 & 0 & 1 & 1 & 1 & 1 \end{pmatrix} \begin{matrix} \text{Carl's Dog} \\ \text{Breed 1} \\ \text{Breed 2} \\ \text{Breed 3} \\ \text{Breed 4} \\ \text{Breed 5} \end{matrix} .$$

a. Use PCA, limited to the two largest singular values, to plot a representation of these six dog types. Be sure to label all of the dog types on the graph.

b. Use your graph from part a. to argue which breed or breed(s) are closest genetically to Carl's dog.

6. Table 3.4 shows Hectares of Palm Oil Plantations in Indonesia versus Sumatran Orangutan populations.

TABLE 3.4
Orangutan Population versus Palm Oil Plantations.

Year	Total Hectares of Palm Oil Plantations	Sumatran Orangutan Population in Indonesia	Population of Indonesia
2000	4,158,077	13,500	211,513,823
2001	4,713,431	11,245	214,427,417
2002	5,067,058	10,254	217,357,793
2003	5,283,557	8,700	220,309,469
2004	5,566,635	7,500	223,285,676
2005	5,950,349	7,200	226,289,470
2006	6,250,460	6,000	229,318,262

a. Use PCA to determine a linear model for to predict the orangutan population based on the number of hectares of palm oil plantations.

b. Use Principal Component Regression to determine a linear model to predict the orangutan population based on both the number of hectares of palm oil plantations and the population of Indonesia.

7. Data related to traits of 86 star husband tales can be found at Github link 23. Use Principal Component Analysis to visualize the 86 tales in $\mathbb{R}^2$ to determine how close the tales are one to another.

8. Data related to 31 Cinderella tales can be found at GitHub link 1. Use Principal Component Analysis to visualize the Cinderella tales in $\mathbb{R}^2$ and determine how close the tales are to one another.

9. A list of 10 East Coast Colleges and Universities and their characteristics are listed in Table 3.5.

TABLE 3.5

College and University Data.

Institution	Out of State Tuition	Number of Undergraduates	Acceptance Rate	Graduation Rate
Bucknell University	56092	4900	.31	.90
Davidson College	51447	1843	.20	.93
Florida State University	18786	32812	.50	.80
Georgia Tech University	33020	16049	.23	.85
Lafayette College	53630	2642	.31	.90
University of North Carolina	35170	19117	.24	.90
University of Richmond	52610	3227	.33	.88
University of Virginia	49032	11786	.27	.93
Villanova University	56730	10983	.29	.91
William and Mary University	45272	6377	.34	.90

a. Standardize the data and use PCA to project each college and university into $\mathbb{R}^2$. Which school appears to be the closest to University of Virginia based on this graph?

b. Project the points from part a. onto the x-axis, using this graph, which school appears to be the closest to University of Virginia.

c. Project the points from part a. onto the line $y = \frac{1}{4}x - \frac{1}{2}$, using this graph, which school appears to be the closest to University of Virginia. Recall that given two vectors u and v, the projection of v onto u

$$proj_u v = \frac{u \cdot v}{v \cdot v} u.$$

d. If data scientists believe that the colleges represented in Table 3.5 should be split into two groups, those with an undergraduate population below 10,000 and those below, determine S_B and S_W.

e. Use your results from part d. and Linear Discriminant Analysis to determine the best vector to project the data onto in order to represent this grouping.

10. S&P 500 stock market data from February 2013 through February 2018 can be found at Github link 48.

$$
Cov = \begin{pmatrix}
 & \text{AAL} & \text{ABBV} & \text{CBG} & \text{DAL} & \text{EQR} & \text{UAL} \\
\text{AAL} & 1. & 0.301441 & 0.458315 & 0.96228 & -0.297844 & 0.91458 \\
\text{ABBV} & 0.301441 & 1. & 0.831858 & 0.307424 & -0.919188 & 0.262801 \\
\text{CBG} & 0.458315 & 0.831858 & 1. & 0.377223 & -0.673998 & 0.351462 \\
\text{DAL} & 0.96228 & 0.307424 & 0.377223 & 1. & -0.296989 & 0.900406 \\
\text{EQR} & -0.297844 & -0.919188 & -0.673998 & -0.296989 & 1. & -0.400563 \\
\text{UAL} & 0.91458 & 0.262801 & 0.351462 & 0.900406 & -0.400563 & 1.
\end{pmatrix}
$$

Cov is the covariance matrix for standardized data for six stocks over this period of time. Discuss how you can use the techniques from this chapter to visualize relationships between these stocks.

4

Interpolation

In Chapter 3, we took a look at one type of regression, using Linear Discriminant Analysis. In Chapters 4 and 5, we will focus on mathematical learning techniques for creating regression models that optimize a given cost function.

In this chapter, we will assume that there is an unknown function, $f(x)$, and that you have been provided the values of this function at $n + 1$ distinct points. These points, $x_0 < x_1 < \cdots < x_n$, called the **interpolation points** can be equidistant or not.

Throughout the chapter, we will refer to the **interpolant** or the **interpolating function**, $\hat{f}(x)$, as the function chosen to model the given interpolation points.

There are several thoughts around what makes an ideal interpolating function. Ideally, given a large set of interpolation points, some of the interpolation points are chosen at random to serve as the **training set** of data used to develop the model, while the rest of the data is used as a **testing set** to test the accuracy of the model.

Many interpolants serve only to predict the value at the given interpolation points. The goal of this type of interpolation problem is to find a function, $\hat{f}(x)$, such that

$$\hat{f}(x_j) = f(x_j) \text{ for all } 0 \leq j \leq n.$$

Although very accurate at the given points, these models, such as the Lagrange interpolation functions, and the Hermitian interpolation functions lack flexibility when attempting to extrapolate or predict near the given points but not within the given set.

For more flexibility, learning systems integrate **regularization**, a technique for smoothing across the data. The neural network methods presented in Chapter 5 incorporate optimization techniques called gradient descent with regularization.

Keep in mind that there is not one solution for a best model. Throughout this chapter, we will be discussing advantages and disadvantages of each model. It is important that you analyze your model in context and determine the best model for the situation.

DOI: 10.1201/9781003025672-4

4.1 Lagrange Interpolation

A Lagrange interpolant, $P(x)$, is an nth degree polynomial passing through $n + 1$ known interpolation points, $(x_0, f(x_0))$, $(x_1, f(x_1))$, $\ldots$, $(x_n, f(x_n))$ where $x_0 < x_1 < \cdots < x_n$, with the additional constraint that

$$P(x) = \sum_{j=0}^{n} P_j(x),$$

where,

$$P_j(x) = \prod_{k=0, k \neq j}^{n} f(x_j) \frac{x - x_k}{x_j - x_k}.$$

One can also think of the Lagrange interpolant in terms of Vandermonde determinants.

The **Vandermonde matrix**, V, is a square matrix of the form

$$V = \begin{pmatrix} 1 & x_0 & x_0^2 & \cdots & x_0^n \\ 1 & x_1 & x_1^2 & \cdots & x_1^n \\ 1 & x_2 & x_2^2 & \cdots & x_2^n \\ \vdots & \vdots & \vdots & \vdots \\ 1 & x_n & x_n^2 & \cdots & x_n^n \end{pmatrix}$$

Notice that given $n + 1$ interpolation points, $x_0, x_1, \ldots, x_n$, when creating a nth degree Lagrange interpolating polynomial,

$$P_j(x) = \sum_{j=0}^{n} (-1)^j f(x_j) \frac{|V_j|}{|V^*|},$$

where

$$V^* = \begin{pmatrix} 1 & x_0 & x_0^2 & \cdots & x_0^n \\ 1 & x_1 & x_1^2 & \cdots & x_1^n \\ 1 & x_2 & x_2^2 & \cdots & x_2^n \\ \vdots & \ddots & \ddots & \ddots & \vdots \\ 1 & x_n & x_n^2 & \cdots & x_n^n \end{pmatrix},$$

$$V_0 = \begin{pmatrix} 1 & x & x^2 & \cdots & x^n \\ 1 & x_1 & x_1^2 & \cdots & x_1^n \\ 1 & x_2 & x_2^2 & \cdots & x^n \\ \vdots & \ddots & \ddots & \ddots & \vdots \\ 1 & x_n & x_n^2 & \cdots & x_n^n \end{pmatrix},$$

$$V_1 = \begin{pmatrix} 1 & x & x^2 & \cdots & x^n \\ 1 & x_0 & x_0^2 & \cdots & x_0^n \\ 1 & x_2 & x_2^2 & \cdots & x_2^n \\ \vdots & \ddots & \ddots & \ddots & \vdots \\ 1 & x_n & x_n^2 & \cdots & x_n^n \end{pmatrix},$$

and in general

$$V_k = \begin{pmatrix} 1 & x & x^2 & \cdots & x^n \\ 1 & x_0 & x_0^2 & \cdots & x_0^n \\ \vdots & \ddots & \ddots & \ddots & \vdots \\ 1 & x_{k-1} & x_{k-1}^2 & \cdots & x_{k-1}^n \\ 1 & x_{k+1} & x_{k+1}^2 & \cdots & x_{k+1}^n \\ \vdots & \ddots & \ddots & \ddots & \vdots \\ 1 & x_n & x_n^2 & \cdots & x_n^n \end{pmatrix}.$$

Notice that in V_k the x_k is not included in the matrix.

Example 4.1. *We will find the unique degree three Lagrange polynomial, $P_3(x)$, agreeing with the data $\{(-1,3), (0,-4), (1,5), (2,-6)\}$.*

$$V^* = \begin{pmatrix} 1 & -1 & 1 & -1 \\ 1 & 0 & 0 & 0 \\ 1 & 1 & 1 & 1 \\ 1 & 2 & 4 & 8 \end{pmatrix},$$

$$V_0 = \begin{pmatrix} 1 & x & x^2 & x^3 \\ 1 & 0 & 0 & 0 \\ 1 & 1 & 1 & 1 \\ 1 & 2 & 4 & 8 \end{pmatrix}, \quad V_1 = \begin{pmatrix} 1 & x & x^2 & x^3 \\ 1 & -1 & 1 & -1 \\ 1 & 1 & 1 & 1 \\ 1 & 2 & 4 & 8 \end{pmatrix},$$

$$V_2 = \begin{pmatrix} 1 & x & x^2 & x^3 \\ 1 & -1 & 1 & -1 \\ 1 & 0 & 0 & 0 \\ 1 & 2 & 4 & 8 \end{pmatrix}, \quad V_3 = \begin{pmatrix} 1 & x & x^2 & x^3 \\ 1 & -1 & 1 & -1 \\ 1 & 0 & 0 & 0 \\ 1 & 1 & 1 & 1 \end{pmatrix}.$$

Then,

$$P_3(x) = f(x_0)\frac{|V_0|}{|V^*|} - f(x_1)\frac{|V_1|}{|V^*|} + f(x_2)\frac{|V_2|}{|V^*|} - f(x_3)\frac{|V_3|}{|V^*|}$$

$$= -\frac{1}{2}x\left(x^2 - 3x + 2\right) - \left(4 - 2x - 4x^2 + 2x^3\right)$$

$$\quad - \frac{5}{2}\left(x^3 - x^2 - 2x\right) + \left(x - x^3\right)$$

$$= -4 + 7x + 8x^2 - 6x^3.$$

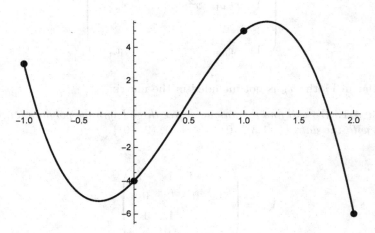

FIGURE 4.1
$P_3(x)$ through the points $\{(-1,3), (0,-4), (1,5), (2,-6)\}$.

4.2 Orthogonal Families of Polynomials

Several of the interpolation techniques presented in this chapter depend on special families of functions. This section focuses on the connection between these special families and linear algebra concepts.

In Section 3.1, we reviewed the concept of an inner product. We see those concepts applied here to families of functions.

Example 4.2. *Let V be defined as the set of integrable functions such that if f and g are in V,*

$$< f, g >= \int_0^{2\pi} f(x)g(x)dx.$$

Let $S = \{1, \ cos(x), \ sin(x)\}$, since

$$\int_0^{2\pi} cos(x)dx = \int_0^{2\pi} sin(x)dx = \int_0^{2\pi} cos(x)sin(x)dx = 0,$$

we say that the set S is orthogonal. Also note that

$$\int_0^{2\pi} cos^2(x)dx = \int_0^{2\pi} sin^2(x)dx = \int_0^{2\pi} 1 \ dx = 1,$$

and thus the set S is orthonormal.

There are a variety of families of orthogonal functions that are generated from matrices of special forms.

Square matrices of the form

$$\begin{pmatrix} x & 1 & 0 & \cdots & 0 & 0 & 0 \\ 1 & 2x & 1 & 0 & \cdots & 0 & 0 \\ 0 & 1 & 2x & 1 & 0 & \cdots & 0 \\ \vdots & \ddots & \ddots & \ddots & \ddots & \vdots & \vdots \\ 0 & \cdots & 0 & 0 & 0 & 1 & 2x \end{pmatrix}$$

are a special type of **tridiagonal matrix**, whose determinants can be viewed recursively as

$$T_{n+1}(x) = 2xT_n - T_{n-1},$$

$T_0(x) = 1$, $T_1(x) = x$, and $T_2(x) = 2x^2 - 1$.

This family of polynomials is called the **Chebyshev polynomials** of the first kind and is orthogonal when we restrict to the domain to $[-1, 1]$ with the defined inner product

$$< f(x), g(x) >= \int_{-1}^{1} (1 - x^2)^{-1/2} f(x)g(x)dx.$$

The Chebyshev polynomials of the first kind can also be defined by

$$T_n(x) = \sum_{m=0}^{\lfloor n/2 \rfloor} \binom{n}{2m} x^{n-2m} (x^2 - 1)^m.$$

Example 4.3. $T_3(x) = (4x^3 - 3x)$. *Note that*

$$\int_{-1}^{1} T_2(x)T_1(x)dx = \int_{-1}^{1} (1 - x^2)^{-1/2}(2x^2 - 1) \cdot x \ dx = 0,$$

$$\int_{-1}^{1} T_3(x)T_1(x)dx = \int_{-1}^{1} (1 - x^2)^{-1/2}(4x^3 - 3x) \cdot x \ dx = 0,$$

$$\int_{-1}^{1} T_3(x)T_2(x)dx = \int_{-1}^{1} (1 - x^2)^{-1/2}(4x^3 - 3x) \cdot (2x^2 - 1) \ dx = 0.$$

Similar relationships can be seen between each pair of Chebyshev polynomials of the first kind. However, the individual Chebyshev polynomials do not have a norm of length 1,

$$\int_{-1}^{1} T_2(x)T_2(x)dx = \int_{-1}^{1} (1 - x^2)^{-1/2}(2x^2 - 1)^2 dx = \frac{\pi}{2}.$$

The set of Chebyshev polynomials of the first kind as defined under the inner product presented here is orthogonal but not orthonormal.

Like the Chebyshev polynomials of the first kind, the **Hermite polynomials** are another family of orthogonal functions that can be generated by matrices of a special form. The Hermite polynomials are also an important family of functions that arise in probability theory.

To take a closer look at Hermite polynomials, we look at derivatives of the polynomial

$$y = a_0 + a_1 x + \cdots + a_n x^n.$$

If

$$A = \begin{pmatrix} a_0 \\ a_1 \\ \vdots \\ a_n \end{pmatrix},$$

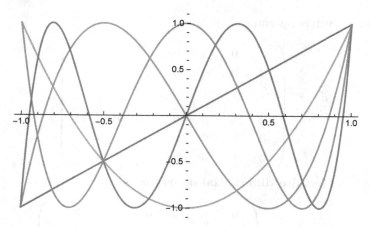

FIGURE 4.2
The first 5 Chebyshev polynomials of the first kind on $[-1, 1]$.

then we denote

$$\frac{dA}{dx} = \begin{pmatrix} a_1 \\ 2a_2 \\ \vdots \\ na_{n-1} \\ 0 \end{pmatrix}$$

and

$$\frac{d^2 A}{dx^2} = \begin{pmatrix} 2a_2 \\ 6a_3 \\ \vdots \\ n(n-1)a_{n-2} \\ 0 \\ 0 \end{pmatrix}.$$

Using these ideas, we introduce the first derivative operator of A as,

$$\frac{d}{dx} \to \begin{pmatrix} 0 & 1 & 0 & 0 & \dots & 0 \\ 0 & 0 & 2 & 0 & \dots & 0 \\ \vdots & \ddots & \ddots & \ddots & \ddots & \vdots \\ 0 & 0 & 0 & 0 & \dots & n \end{pmatrix},$$

the second derivative operator of A as

$$\frac{d^2}{d^2x} \rightarrow \begin{pmatrix} 0 & 0 & 2 & 0 & \cdots & 0 \\ 0 & 0 & 0 & 6 & \cdots & 0 \\ \ddots & \ddots & \ddots & \ddots & \ddots & \vdots \\ 0 & 0 & 0 & 0 & \cdots & n(n-1) \\ 0 & 0 & 0 & 0 & \cdots & 0 \\ 0 & 0 & 0 & 0 & \cdots & 0 \end{pmatrix},$$

and the $n \times n$ **Hermite differential operator** as

$$\frac{d^2}{d^2x} - 2x\frac{d}{dx} \rightarrow \begin{pmatrix} 0 & 0 & 2 & 0 & \cdots & 0 \\ 0 & -2 & 0 & 6 & \cdots & 0 \\ 0 & 0 & -4 & 0 & \cdots & 0 \\ \cdots & \vdots & \vdots & \vdots & \vdots & \cdots \\ 0 & 0 & 0 & 0 & \cdots & n(n-1) \\ 0 & 0 & 0 & 0 & \cdots & 0 \\ 0 & 0 & 0 & 0 & \cdots & -2n \end{pmatrix}.$$

If $\lambda_0 \geq \lambda_1 \geq \cdots \geq \lambda_n$ are the eigenvalues of the $n \times n$ differential operator $\frac{d^2}{d^2x} - 2x\frac{d}{dx}$ with corresponding eigenvectors $v_0, v_1, \ldots, v_n$ then the coefficients of the kth Hermite polynomial corresponds to the eigenvector v_{2k}.

Example 4.4. *The 5×5 Hermite differential operator corresponds to the matrix*

$$\begin{pmatrix} 0 & 0 & 2 & 0 & 0 \\ 0 & -2 & 0 & 6 & 0 \\ 0 & 0 & -4 & 0 & 12 \\ 0 & 0 & 0 & -6 & 0 \\ 0 & 0 & 0 & 0 & -8 \end{pmatrix}$$

which has eigenvectors and corresponding Hermite polynomials

$$
\begin{aligned}
v_1 &= (1,0,0,0,0), & H_1(x) &= 1 \\
v_2 &= (0,2,0,0,0), & H_2(x) &= 2x \\
v_3 &= (-2,0,4,0,0), & H_3(x) &= 4x^2 - 2 \\
v_4 &= (0,-12,0,8,0), & H_4(x) &= 8x^3 - 12x \\
v_5 &= (12,0,-48,0,16), & H_5(x) &= 16x^4 - 48x^2 + 12
\end{aligned}
$$

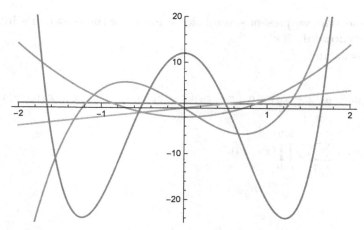

FIGURE 4.3
The first 5 Hermite polynomials on $[-2, 2]$.

The Hermite polynomials can also be generated using the recursive definition

$$H_n(x) = xH_{n-1}(x) - nH_{n-2}(x),$$

Under the inner product defined by

$$< f(x), \, g(x) > = \int_{-\infty}^{\infty} e^{-x^2} f(x)g(x) \ dx,$$

the Hermite polynomials, like the Chebyshev polynomials of the first kind, are an orthogonal set of functions.

Both the Chebyshev polynomials of the first kind and the Hermite polynomials are popular families of functions used for interpolating, or curve fitting, of data.

4.3 Newton's Divided Difference

Section 4.1 presents Lagrange interpolating polynomials which fit exactly to a given set of interpolating points. In fact, given an unknown function $f(x)$ and $n + 1$ interpolating points, $x_0 < x_1 < \cdots < x_n$, there is a unique polynomial $\hat{f}(x)$ such that

$$\widehat{f}(x_j) = f(x_j), \, 0 < j < n.$$

In this section, we present several closed forms for the **Newton's interpolating polynomial,** $\widehat{f}(x)$.

Let $\widehat{f}_1(x) = a_0$ and

$$\widehat{f}_n(x) = a_0 + a_1(x - x_0) + a_2(x - x_0)(x - x_1) + \cdots + a_n(x - x_0)(x - x_1) \cdots$$
$$(x - x_n)$$
$$= a_0 + \sum_{j=1}^{n} a_j \prod_{i=0}^{j-1} (x - x_i),$$

where

$$a_j = \frac{f(x_j) - \widehat{f}_{j-1}(x_j)}{\prod_{i=0}^{j-1}(x_j - x_i)}.$$

You might have noticed that Newton's interpolating polynomials are defined recursively. That is, that the definition of the nth degree polynomial, $\widehat{f}_n(x)$, depends on $\widehat{f}_1(x)$, ..., $\widehat{f}_{n-1}(x)$.

We can also write Newton's interpolating polynomials, recursively, in terms of **Newton's divided differences.**

4.3.1   Newton's Interpolation via Divided Difference

The first **Newton's divided difference** between x_i and x_j, denoted

$$f[x_i, x_j],$$

is the slope of the tangent line between $(x_i, f(x_i))$ and $(x_j, f(x_j))$,

$$f[x_i, x_j] = \frac{f(x_i) - f(x_j)}{x_i - x_j}.$$

Similarly, $n + 1$ points $x_0 < x_1 < \cdots < x_n$ can be used to find the second divided difference,

$$f[x_0, x_1, \ldots, x_n] = \frac{f[x_0, x_1, \ldots, x_{n-1}] - f[x_1, x_2, \ldots, x_n]}{x_0 - x_n}.$$

Typically, Newton's divided differences are presented in a table similar to the one seen in Table 4.1.

TABLE 4.1

Divided Difference where $\Delta x = 1$.

$f[x_0] = 3$			
	$f[x_0,x_1] = -1$		
		$f[x_0,x_1,x_2] = \frac{-3+1}{2} = -1$	
$f[x_1] = 2$			
	$f[x_1,x_2] = -3$		$f[x_0,x_1,x_2,x_3] = \frac{4.5+1}{3}$
$f[x_2] = -1$			
		$f[x_1,x_2,x_3] = \frac{6+3}{2} = 4.5$	
	$f[x_3,x_2] = 6$		
$f[x_3] = 5$			

Applying the concepts of divided difference, Newton's interpolating polynomials can be written as

$$\widehat{f_1}(x) = a_0 + a_1(x - x_0) = f(x_0) + (x - x_0)f[x_0,x_1]$$
$$\widehat{f_2}(x) = a_0 + (x - x_0)f[x_0,x_1] + (x - x_0)(x - x_1)f[x_0,x_1,x_2]$$
$$\widehat{f_3}(x) = a_0 + (x - x_0)f[x_0,x_1] + (x - x_0)(x - x_1)f[x_0,x_1,x_2]$$
$$+ (x - x_0)(x - x_1)(x - x_2)f[x_0,x_1,x_2,x_3]$$

Example 4.5. *Given a set of four interpolating points* $\{(0,3), (1,2), (2,-1), (3,5)\}$,

$$\widehat{f_1}(x) = 3 - (x - 0) = 3 - x$$
$$\widehat{f_2}(x) = 3 - (x - 0) - 1(x - 0)(x - 1) = 3 + x - 2x^2$$
$$\widehat{f_3}(x) = 3 - (x - 0) - 1(x - 0)(x - 1) + \frac{6.5}{3}(x - 0)(x - 1)(x - 2)$$
$$= \frac{1}{6}\left(11x^3 - 39x^2 + 22x + 18\right)$$

Notice that the Newton's interpolating polynomials can be written as a linear system as well.

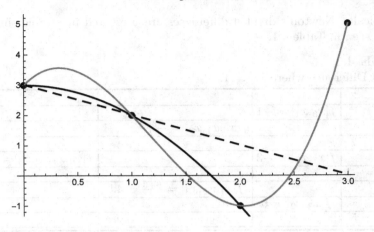

FIGURE 4.4
Data with $\widehat{f}_1(x)$, $\widehat{f}_2(x)$, $\widehat{f}_3(x)$ from Example 4.5.

$$
\begin{pmatrix}
\hat{f}_0(x) \\
\hat{f}_1(x) \\
\hat{f}_2(x) \\
\hat{f}_3(x)
\end{pmatrix}
$$

$$
=
\begin{pmatrix}
1 & 0 & 0 & 0 \\
1 & (x-x_0) & 0 & 0 \\
1 & (x-x_0) & (x-x_0)(x-x_1) & 0 \\
1 & (x-x_0) & (x-x_0)(x-x_1) & (x-x_0)(x-x_1)(x-x_2)
\end{pmatrix}
$$

$$
\begin{pmatrix}
f[x_0] \\
f[x_0,x_1] \\
f[x_0,x_1,x_2] \\
f[x_0,x_1,x_2,x_3]
\end{pmatrix}
$$

4.3.2   Newton's Interpolation via the Vandermonde Matrix

We can also write Newton's divided differences in terms of the Vandermonde matrix. In order to do so, we introduce the concept of LU-decomposition so that we can later decompose the Vandermonde matrix in this context.

Let A be a square matrix. If A can be written in terms of a lower triangular matrix, L, with diagonal entries equal to 1, and an upper triangular matrix, U, such that

$$A = LU$$

then we say that A has a **LU-decomposition**.

If a matrix A has a LU-decomposition, L and U can be determined by finding a sequence of matrices $L_0, L_1, \ldots, L_k$, such that

$$L_k \cdots L_1 L_0 A = U,$$

where L_i is a lower triangular elementary matrix, $0 \leq i \leq k$. Then

$$L = L_0^{-1} \ldots L_k^{-1}.$$

Example 4.6. *Let* $A = \begin{pmatrix} 6 & 18 & 3 \\ 2 & 12 & 1 \\ 4 & 15 & 3 \end{pmatrix}$. *Using Gauss Jordan elimination,*

$$L_0 = \begin{pmatrix} 1 & 0 & 0 \\ -\frac{1}{3} & 1 & 0 \\ 0 & 0 & 1 \end{pmatrix}, L_1 = \begin{pmatrix} 1 & 0 & 0 \\ 0 & 1 & 0 \\ -\frac{2}{3} & 0 & 1 \end{pmatrix}, L_2 = \begin{pmatrix} 1 & 0 & 0 \\ 0 & 1 & 0 \\ 0 & -\frac{1}{2} & 1 \end{pmatrix}.$$

$$A = \begin{pmatrix} 1 & 0 & 0 \\ \frac{1}{3} & 1 & 0 \\ \frac{2}{3} & \frac{1}{2} & 1 \end{pmatrix} \begin{pmatrix} 6 & 18 & 3 \\ 0 & 6 & 0 \\ 0 & 0 & 1 \end{pmatrix}.$$

$L_0, L_1, \ldots, L_k$ *and* U *may not be unique.*

Theorem 11. *If a square matrix A is invertible and has a LU-decomposition, then that decomposition is unique.*

Theorem 12. *Let $S = \{(x_i, f[x_i]) \mid 0 \leq i \leq n-1\}$ be a given set of interpolating points, where the Vandermonde matrix, V, is a square $n \times n$, invertible matrix,*

$$V = \begin{pmatrix} 1 & x_0 & x_0^2 & \cdots & x_0^{n-1} \\ 1 & x_1 & x_1^2 & \cdots & x_1^{n-1} \\ 1 & x_2 & x_2^2 & \cdots & x_2^{n-1} \\ \vdots & \vdots & \vdots & \vdots & \\ 1 & x_{n-1} & x_{n-1}^2 & \cdots & x_{n-1}^{n-1} \end{pmatrix}$$

If $V = LU$ where L is a lower triangular matrix with main diagonal entries equal to 1, and U is an upper triangular matrix, then the elements of L and U respectively are

$$L_{i,1} = L_{i,i} = 1, 1 \leq i \leq n.$$
$$L_{i,j} = x_i, 1 < j < i.$$

Example 4.7. *Given that $x_0 = 0$, $x_1 = 1$, $x_2 = 2$, $x_3 = 3$. The Vandermonde matrix*

$$V = \begin{pmatrix} 1 & 0 & 0 & 0 \\ 1 & 1 & 1 & 1 \\ 1 & 2 & 4 & 8 \\ 1 & 3 & 9 & 27 \end{pmatrix}.$$

Using Gaussian Elimination,

$$L_0 = \begin{pmatrix} 1 & 0 & 0 & 0 \\ -1 & 1 & 0 & 0 \\ 0 & 0 & 1 & 0 \\ 0 & 0 & 0 & 1 \end{pmatrix}, L_1 = \begin{pmatrix} 1 & 0 & 0 & 0 \\ 0 & 1 & 0 & 0 \\ -1 & 0 & 1 & 0 \\ 0 & 0 & 0 & 1 \end{pmatrix},$$

$$L_2 = \begin{pmatrix} 1 & 0 & 0 & 0 \\ 0 & 1 & 0 & 0 \\ 0 & 0 & 1 & 0 \\ -1 & 0 & 0 & 1 \end{pmatrix}, L_3 = \begin{pmatrix} 1 & 0 & 0 & 0 \\ 0 & 1 & 0 & 0 \\ 0 & -2 & 1 & 0 \\ 0 & 0 & 0 & 1 \end{pmatrix}$$

$$L_4 = \begin{pmatrix} 1 & 0 & 0 & 0 \\ 0 & 1 & 0 & 0 \\ 0 & 0 & 1 & 0 \\ 0 & -3 & 0 & 1 \end{pmatrix}, L_5 = \begin{pmatrix} 1 & 0 & 0 & 0 \\ 0 & 1 & 0 & 0 \\ 0 & 0 & 1 & 0 \\ 0 & 0 & -3 & 1 \end{pmatrix}$$

$$L_5 L_4 L_3 L_2 L_1 L_0 V = \begin{pmatrix} 1 & 0 & 0 & 0 \\ 0 & 1 & 1 & 1 \\ 0 & 0 & 2 & 6 \\ 0 & 0 & 0 & 6 \end{pmatrix} = U$$

and

$$L = (L_5 L_4 L_3 L_2 L_1 L_0)^{-1} = \begin{pmatrix} 1 & 0 & 0 & 0 \\ 1 & 1 & 0 & 0 \\ 1 & 2 & 1 & 0 \\ 1 & 3 & 3 & 1 \end{pmatrix}.$$

Note that V is a nonsingular matrix as long as x_i, $0 \leq i \leq n-1$ are unique. We wish to determine coefficients $a = \{a_0, a_1, \ldots, a_n\}$ such that

$$Va = \hat{f}.$$

We can use the LU decomposition of the Vandermonde matrix V to find the divided differences,

$$f[x_0, x_1, \ldots, x_i] = U_{i,i}^{-1} L^{-1} \begin{pmatrix} f[x_0] \\ f[x_1] \\ \vdots \\ f[x_n] \end{pmatrix}.$$

Example 4.8. *Referring back to the data from Table 4.1,*

$$V = \begin{pmatrix} 1 & 0 & 0 & 0 \\ 1 & 1 & 1 & 1 \\ 1 & 2 & 4 & 8 \\ 1 & 3 & 9 & 27 \end{pmatrix}.$$

The LU-decomposition of $V = LU$ produces the matrices

$$L = \begin{pmatrix} 1 & 0 & 0 & 0 \\ 1 & 1 & 0 & 0 \\ 1 & 2 & 1 & 0 \\ 1 & 3 & 3 & 1 \end{pmatrix}$$

$$U = \begin{pmatrix} 1 & 0 & 0 & 0 \\ 0 & 1 & 1 & 1 \\ 0 & 0 & 2 & 6 \\ 0 & 0 & 0 & 6 \end{pmatrix}.$$

The Newton's divided differences can be written in terms of the entries in L^{-1} and U^{-1},

$$L^{-1} = \begin{pmatrix} 1 & 0 & 0 & 0 \\ -1 & 1 & 0 & 0 \\ 1 & -2 & 1 & 0 \\ -1 & 3 & -3 & 1 \end{pmatrix}, U^{-1} = \begin{pmatrix} 1 & 0 & 0 & 0 \\ 0 & 1 & -\frac{1}{2} & \frac{1}{3} \\ 0 & 0 & \frac{1}{2} & -\frac{1}{2} \\ 0 & 0 & 0 & \frac{1}{6} \end{pmatrix}.$$

$$L^{-1} \begin{pmatrix} 3 \\ 2 \\ -1 \\ 5 \end{pmatrix} = \begin{pmatrix} 3 \\ -1 \\ -2 \\ 11 \end{pmatrix}$$

and thus

$$\begin{pmatrix} f[x_0] \\ f[x_0,x_1] \\ f[x_0,x_1,x_2] \\ f[x_0,x_1,x_2,x_3] \end{pmatrix} = \begin{pmatrix} 3 \\ -1 \\ -1 \\ \frac{11}{6} \end{pmatrix}$$

4.4   Chebyshev Interpolation

For the interpolating techniques presented in this chapter thus far, we have assumed that interpolating points are given. There are instances where you are able to choose the interpolating points or that equally spaced interpolating points can be transformed onto another set of points.

Chebyshev interpolation is a method when n interpolating points are located at the **Chebyshev nodes**, the roots of the nth degree Chebyshev polynomial of the first kind, $T_n(x)$,

$$x_k = cos\left(\frac{(2k+1)\pi}{2n}\right), k = 0, 1, 2, \ldots, n-1.$$

An nth-degree polynomial, $p(x)$, interpolating a function $f(x)$, can be written as the sum of Chebyshev polynomials of the first kind,

$$p(x) = \frac{1}{2}c_0 T_0(x) + \sum_{i=1}^{n} c_i T_i(x),$$

where

$$c_i = \frac{2}{n} \sum_{k=0}^{n} f(x_k)T_i(x_k).$$

Example 4.9. *Assume that you have a dataset, S, of 4 equidistant points generated from the function*

$$f(x) = \frac{1}{1 + x^2}$$

over the interval $[-1, 1]$. *We will find a 3rd-degree Chebyshev interpolating polynomial, $p(x)$.*

Since we are to find a 3rd-degree interpolating polynomial, we will use four interpolating points, which are roots of

$$T_4(x) = 8x^4 - 8x^2 + 1,$$

These roots are

$$x_0 = -\frac{\left(\sqrt{2 + \sqrt{2}}\right)}{2}, \; x_1 = -\frac{\left(\sqrt{2 - \sqrt{2}}\right)}{2}, \; x_2 = \frac{\left(\sqrt{2 - \sqrt{2}}\right)}{2}, \; x_3 = \frac{\left(\sqrt{2 + \sqrt{2}}\right)}{2}.$$

The corresponding values of $f(x)$ are

$$f(x_0) = f(x_3) = \frac{4}{(6 + \sqrt{2})},$$

$$f(x_1) = f(x_2) = \frac{2(6 + \sqrt{2})}{17}.$$

Recall that $T_0(x) = 1$, $T_1(x) = x$, $T_2(x) = 2x^2 - 1$, and $T_3(x) = 4x^3 - 3x$ and thus the generating coefficients

$$c_0 = \frac{2}{4} \sum_{i=0}^{3} f(x_i) T_0(x_i) \approx 1.41176$$

$$c_1 = \frac{2}{4} \sum_{i=0}^{3} f(x_i) T_1(x_i) = 0$$

$$c_2 = \frac{2}{4} \sum_{i=0}^{3} f(x_i) T_2(x_i) \approx -0.235294$$

$$c_3 = \frac{2}{4} \sum_{i=0}^{3} f(x_i) T_3(x_i) = 0$$

The interpolating polynomial is

$$p(x) = \frac{1}{2} c_0 T_0(x) + \sum_{i=1}^{n} c_i T_i(x) = 0.705882 - 0.235294 \left(2x^2 - 1\right).$$

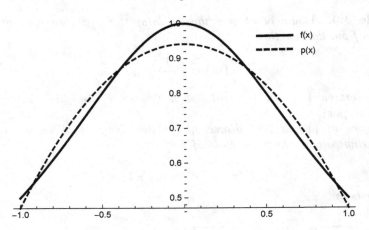

FIGURE 4.5
Example of Chebyshev interpolation of $f(x) = \frac{1}{1+x^2}$.

It is important to note that Chebyshev interpolation is traditionally meant to be conducted on the interval $[-1, 1]$ since the roots of the Chebyshev polynomials are on the unit circle. Thus, if you intended to use Chebyshev interpolation on an interval $[a, b]$ you must first create a linear transformation

$$\mathcal{T} : [-1, 1] \to [a, b], \text{ where } u(-1) = a \text{ and } u(1) = b.$$

If $u = \alpha x + \beta$ then

$$a = -\alpha + \beta, b = \alpha + \beta$$

and $u = \frac{b-a}{2}x + \frac{b+a}{2}$.

Example 4.10. *Let's say that you would like to approximate the function* $f(x) = e^x$ *on the interval* $[-1, 3]$ *using a* 5th*-degree Chebyshev interpolation. Using* $T_0(x)$, $T_1(x)$, ..., *and* $T_4(x)$ *seen previously in Example 4.9,*

$$T_5(x) = 16x^5 - 20x^3 + 5x \text{ and } T_6(x) = 32x^6 - 48x^4 + 18x^2 - 1.$$

$T_6(x)$ *has roots*

$$x_k = \cos\left(\frac{(2k+1)\pi}{12}\right), k = 0, 1, \ldots, 5.$$

Roots on the interval $[-1, 3]$ *can be mapped to the interval* $[-1, 1]$ *using the linear transformation* $u = 2x + 1$. *Under this transformation,* $u_k = 2x_k + 1$ *and* $c_i = \frac{2}{6} \sum_{i=0}^{n-1} e^{\frac{u_k-1}{2}} T_i(x_k)$.

$$c_0 = \frac{2}{4} \sum_{i=0}^{3} f(u_i) T_0(x_i) \approx 6.19656$$

$$c_1 = \frac{2}{4} \sum_{i=0}^{3} f(u_i) T_1(x_i) \approx 8.6476$$

$$c_2 = \frac{2}{4} \sum_{i=0}^{3} f(u_i) T_2(x_i) \approx 3.74551$$

$$c_3 = \frac{2}{4} \sum_{i=0}^{3} f(u_i) T_3(x_i) \approx 1.15656$$

$$c_4 = \frac{2}{4} \sum_{i=0}^{3} f(u_i) T_4(x_i) \approx 0.275639$$

$$c_5 = \frac{2}{4} \sum_{i=0}^{3} f(u_i) T_5(x_i) \approx 0.000539728$$

Using these coefficients,

$$p(x) = \frac{1}{2} c_0 T_0(u) + \sum_{i=1}^{5} c_i T_i(u)$$

$$= 0.0002699x^5 + 0.1365x^4 + 0.02835x^3 + 0.4149x^2 + 1.128x + 1.019$$

It may seem obscure to be able to choose interpolation points at the Chebyshev nodes; however, Chebyshev interpolation is used extensively in numerical analysis, particularly in creating filters, due to it's ability to minimize **Runge's phenomenon**.

Runge's phenomenon occurs with polynomial interpolants of high degree over a set of equally spaced interpolation points. Runge's phenomenon is when the interpolant matches the desired function or data extremely well in the middle of the interpolation interval and begins to oscillate at the edges of the interval. This phenomena should be considered when deciding to increase the degree of your polynomial interpolant. An example of Runge's phenomenon can be seen in Figure 4.7.

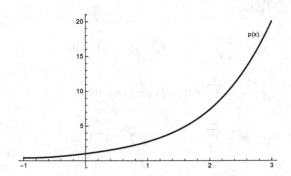

FIGURE 4.6
Example of Chebyshev interpolation of $f(x) = e^x$ on $[-1, 3]$.

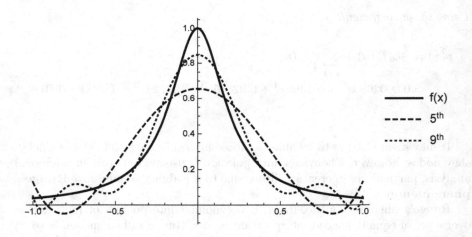

FIGURE 4.7
Example of Runge's phenomenon with $f(x) = \frac{1}{1+25x^2}$ on $[-1, 1]$ with 5th and 9th degree interpolants.

4.5 Hermite Interpolation

Thus far we have been focusing on creating an interpolating function based just on interpolating points. **Hermite interpolation** constructs a interpolating function that involves both function and derivative values.

Like Lagrange interpolation, a hermite interpolating function, $h(x)$, requires that, for an unknown function $f(x)$, $h(x)$ matches $f(x)$ and its first p derivatives on a set of $n + 1$ data points. That is,

$$h(x_i) = f(x_i), \text{ for } 0 \leq i \leq n$$
$$h^{(k)}(x_i) = f^{(k)}(x_i), \text{ for } 0 \leq i \leq n, 0 \leq k \leq p.$$

By definition $h(x)$ has $(n+1)(p+1)$ constraints imposed on it and thus we will be seeking a polynomial of $(n+1)(p+1) - 1$ degree to satisfy all of these requirements.

We begin with an example before making general statements around hermite polynomials.

Example 4.11. *Given $f(0) = 1$, $f(1) = 6$, $f'(0) = 2$, $f'(1) = 12$, $f''(0) = 0$, and $f''(1) = 24$. From this given information, we find the hermite interpolating function*

$$h(x) = a_0 + a_1 x + a_2 x^2 + a_3 x^3 + a_4 x^4 + a_5 x^5.$$

Notice that

$$h(0) = a_0 = 1$$
$$h'(0) = a_1 = 2$$
$$h''(0) = 2a_2 = 0$$
$$h(1) = a_0 + a_1 + a_2 + a_3 + a_4 + a_5 = 6$$
$$h'(1) = a_1 + 2a_2 + 3a_3 + 4a_4 + 5a_5 = 12$$
$$h''(1) = 2a_2 + 6a_3 + 12a_4 + 20a_5 = 24.$$

This system can be written in the form $Ax = b$,

$$\begin{pmatrix} 1 & 0 & 0 & 0 & 0 & 0 \\ 0 & 1 & 0 & 0 & 0 & 0 \\ 0 & 0 & 2 & 0 & 0 & 0 \\ 1 & 1 & 1 & 1 & 1 & 1 \\ 0 & 1 & 2 & 3 & 4 & 5 \\ 0 & 0 & 2 & 6 & 12 & 20 \end{pmatrix} \begin{pmatrix} a_0 \\ a_1 \\ a_2 \\ a_3 \\ a_4 \\ a_5 \end{pmatrix} = \begin{pmatrix} 1 \\ 2 \\ 0 \\ 6 \\ 12 \\ 24 \end{pmatrix}.$$

If A is the coefficient matrix from the above system, then we can again solve for a_0, a_1, ..., a_5 by determining $(AA^T)^{-1}A^T b$,

$$h(x) = 1 + 2x + 2x^3 + x^4.$$

If we wish to concern ourselves with exploring only those Hermite interpolating functions that satisfy conditions up through the first derivative, $h(x_i) = f(x_i)$ and $h'(x_i) = f'(x_i)$, we can write such Hermite interpolating functions in terms of Newton's divided difference if the known points $(x_i, f'(x_i))$ are treated as interpolating points as well.

TABLE 4.2
Divided Difference Incorporating the First Derivative.

z	$f[z]$	$f[z_i,z_{i+1}]$	$f[z_i,z_{i+1},z_{i+2}]$	$f[z_i,z_{i+1},z_{i+2},z_{i+3}]$
$z_0 = x_0 = 0$	1			
		$f[z_0,z_1] = f'[x_0] = 2$		
$z_1 = 0$	1			
			$f[z_0,z_1,z_2] = 4$	
		$f[z_1,z_2] = \frac{f[z_1]-f[z_2]}{z_1-z_2} = 6$		
				$f[z_0,z_1,z_2,z_3] = 2$
$z_2 = x_1 = 1$	6			
			$f[z_1,z_2,z_3] = 6$	
		$f[z_2,z_3] = f'[x_1] = 12$		
$z_3 = 1$	6			

Example 4.12. *Revisiting Example 4.5 and using Table 4.2 to provide a Hermite interpolating polynomial that matches both the function values and first derivative values, $h(x)$ is a third degree polynomial.*

$$\begin{aligned}
h(z) &= a_0 + (z - z_0)f[z_0,z_1] + (z - z_0)(z - z_1)f[z_0,z_1,z_2] \\
&\quad + (z - z_0)(z - z_1)(z - z_2)f[z_0,z_1,z_2,z_3] \\
&= 0 + 2(z - 0) + 4(z - 0)(z - 0) + 2(z - 0)(z - 0)(z - 1) \\
&= 2z^3 + 2z^2 + 2z
\end{aligned}$$

4.6   Least Squares Regression

This section focuses on determining interpolating polynomials satisfying

$$\hat{f}_n(x) = \sum_{i=0}^{n} a_i x^n, \ \hat{f}(x_j) = f(x_j), \ 0 \le j \le n.$$

that minimize the sum of squared errors, $SSE = \sum_{j=0}^{n}(\hat{f}(x_j) - f(x_j))^2$. The above system can be written as a system of linear equations,

$$\begin{pmatrix} 1 & x_0 & \cdots & x_0^n \\ 1 & x_1 & \cdots & x_1^n \\ \vdots & \vdots & \vdots & \vdots \\ 1 & x_n & \cdots & x_n^n \end{pmatrix} \begin{pmatrix} a_0 \\ a_1 \\ \vdots \\ a_n \end{pmatrix} = \begin{pmatrix} f(x_0) \\ f(x_1) \\ \vdots \\ f(x_n) \end{pmatrix}.$$

In order for this system to have a unique solution, the **Vandermonde determinant**

$$\begin{vmatrix} 1 & x_0 & \cdots & x_0^n \\ 1 & x_1 & \cdots & x_1^n \\ \vdots & \vdots & \vdots & \vdots \\ 1 & x_n & \cdots & x_n^n \end{vmatrix} = \prod_{i>j}(x_i - x_j) \neq 0.$$

Additionally, one might wish to fit an mth degree interpolating polynomial to $n + 1$ given interpolating points, where $m < n + 1$.

In this case, the matrix

$$A = \begin{pmatrix} 1 & x_0 & \cdots & x_0^m \\ 1 & x_1 & \cdots & x_1^m \\ \vdots & \vdots & \vdots & \vdots \\ 1 & x_n & \cdots & x_n^m \end{pmatrix}$$

is not square, so is definitely not invertible. Fortunately, the system is still of the form $Ax = b$, where

$$x = \begin{pmatrix} a_0 \\ a_1 \\ \vdots \\ a_m \end{pmatrix} \text{ and } b = \begin{pmatrix} f(x_0) \\ f(x_1) \\ \vdots \\ f(x_n) \end{pmatrix}.$$

Note that $A^T A$ is an $m \times m$ matrix and $A^T A x = A^T b$. Thus, we can solve for the coefficient vector

$$x = (A^T A)^{-1} A^T b.$$

So how do we know that this solution for x is the solution that minimizes the SSE?

Recall that the entries of b are the values on the interpolation function. The residual errors,

$$\vec{r} = proj_{Ax} b.$$

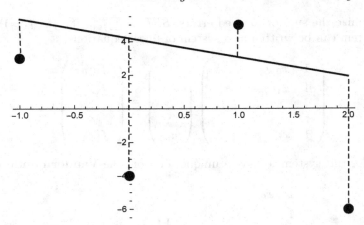

FIGURE 4.8
An interpolant and the residuals.

A visualization of this can be seen in Figure 4.8.

Recall from Linear Algebra that if $Ax = b$, then b is in the range of the linear transformation Ax, which also means that b is in the columnspace of A. Additionally, $\vec{r}$ is orthogonal to the columnspace of A.

$$\vec{r} = b - Ax$$
$$A^T \cdot \vec{r} = A^T(b - Ax) = 0.$$

Solving for x, we see that

$$A^T b - A^T Ax = 0,$$
$$x = (A^T A)^{-1} A^T b.$$

Example 4.13 shows us how to apply the least squares regression technique in a real world example.

Example 4.13. *Some might claim that the number of total COVID 19 cases in North Carolina in the months from July 2020 to December 2020 grew exponentially.*

In this example, we begin by using least squares regression to fit a quadratic interpolating polynomial to the data in Table 4.3 and will conclude with a discussion of how to fit exponential interpolating functions using similar techniques.

TABLE 4.3

COVID 19 total cases in NC in 2020.

Date	July 1	Aug. 1	Sept. 1	Oct. 1	Nov. 1	Dec 1
Total Infections	65397	124835	172293	212909	276692	367000

The coefficients of a least squares quadratic interpolating function $f(x) = a_0 + a_1x + a_2x^2$ to fit the data in Table 4.3 can be found by solving

$$\begin{pmatrix} a_0 \\ a_1 \\ a_2 \end{pmatrix} = \begin{pmatrix} 6 & 21 & 91 \\ 21 & 91 & 441 \\ 91 & 441 & 2275 \end{pmatrix}^{-1} \begin{pmatrix} 1 & 1 & 1 & 1 & 1 & 1 \\ 1 & 2 & 3 & 4 & 5 & 6 \\ 1 & 4 & 9 & 16 & 25 & 36 \end{pmatrix} \begin{pmatrix} 65397 \\ 124835 \\ 172293 \\ 212909 \\ 276692 \\ 367000 \end{pmatrix}$$

$$= \begin{pmatrix} 39375.8 & 29806.7 & 3922.32 \end{pmatrix}.$$

If instead, one wishes to fit the least squares interpolating function

$$g(x) = e^{a_0 + a_1 x}$$

to the given data. This is equivalent to fitting the interpolating function

$$ln(g(x)) = a_0 + a_1x, \text{ where}$$

$$\begin{pmatrix} a_0 \\ a_1 \end{pmatrix} = (A^T A)^{-1} A^T ln \begin{pmatrix} 65397 \\ 124835 \\ 172293 \\ 212909 \\ 276692 \\ 367000 \end{pmatrix}$$

$$= \begin{pmatrix} \frac{13}{15} & -\frac{1}{5} \\ -\frac{1}{5} & \frac{2}{35} \end{pmatrix} \begin{pmatrix} 1 & 1 & 1 & 1 & 1 & 1 \\ 1 & 2 & 3 & 4 & 5 & 6 \end{pmatrix} ln \begin{pmatrix} 65397 \\ 124835 \\ 172293 \\ 212909 \\ 276692 \\ 367000 \end{pmatrix}$$

$$\begin{pmatrix} a_0 \\ a_1 \end{pmatrix} = \begin{pmatrix} 10.9597 \\ 0.320681 \end{pmatrix}.$$

So $ln(g(x)) = 10.9597 + 0.320681x$ and $g(x) = e^{10.9597 + 0.320681x}$.

The two interpolants determined in this example can be seen in Figure 4.9.

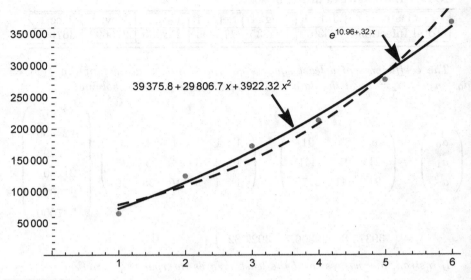

FIGURE 4.9
Interpolating function for the data in Example 4.13.

There are several ways to think about the error of an interpolant, although the goal of least square regression is to minimize the sum of the squared errors **SSE**,

$$SSE = (l_2 \text{ error})^2 = \sum_{j=o}^{n} (\hat{f}(x_j) - f(x_j))^2.$$

Additional errors to consider are the **absolute error**, which is equal to

$$\sum_{j=o}^{n} |\hat{f}(x_j) - f(x_j)|,$$

and the **relative l_2 error**, which is

$$\frac{\sum_{j=o}^{n} \sqrt{(\hat{f}(x_j) - f(x_j))^2}}{\sum_{j=o}^{n} \sqrt{f(x_j)^2}}.$$

Example 4.14. *For the interpolants in Example 4.13, an error analysis can be see in Table 4.4.*

TABLE 4.4
Errors for Example 4.13.

	$39375.8 + 29806.7x + 3922.32x^2$	$e^{10.96+0.32x}$
Absolute error	51867.1	92759.8
l_2 error	21311.5	41785.6
Relative l_2 error	0.0385	0.0755

4.7 CASE STUDY: Chebyshev Polynomials and Cryptography

There are many techniques for sending an encrypted code and decrypting that code. In this case study, we see a few different techniques related to linear algebra and discuss their qualities.

If Special Agents Rahasa and Siri agree to use Hamming encryption, Rahasa would then place her message into a $2 \times n$ matrix. Rahasa begins by testing their communications with the message, "This is a test", which translates to the matrix

$$M = \begin{pmatrix} 20 & 9 & 1 & 5 & 20 \\ 8 & 19 & 20 & 19 & 20 \end{pmatrix}.$$

Notice that each letter is assigned its position in the alphabet, and if Rahasa has an odd number of letters in her message, then she just replicates the last letter. Rahasa then creates a 2×2 encryption matrix, A, which is invertible mod 29.

$$A = \begin{pmatrix} 1 & 2 \\ 3 & 4 \end{pmatrix}$$

$$A^{-1} \bmod 29 \equiv \begin{pmatrix} 27 & 1 \\ 16 & 14 \end{pmatrix},$$

and sends the encryption matrix A as well as the encrypted message affiliated with

$$\widetilde{M} = A.M \bmod 29 \equiv \begin{pmatrix} 7 & 18 & 12 & 14 & 2 \\ 5 & 16 & 25 & 4 & 24 \end{pmatrix}.$$

"GERPLYNDBX."

Given A and the encrypted message, Siri can recreate $\widetilde{M}$ and determine $M \equiv A^{-1}\widetilde{M} \bmod 29$ and the original matrix.

Unfortunately, Rahasa's message might be intercepted by a third party, special agent Gizli. In order to prevent this, Rahasa and Siri have decided to use a more secure method of transmitting their message, using **Public Key Cryptography**.

Public Key Cryptography allows users who do not share any secret key to securely communicate over a public channel.

There are several different public key algorithms. In this case study, we explore a public key algorithm based on chaos theory and Chebyshev polynomials.

In addition to being orthogonal polynomials, Chebyshev polynomials have the remarkable property that

$$T_r(T_s(x)) = T_{rs}(x) = T_s(T_r(x)), \text{ for all } x.$$

For any given prime number, p, it is still true that

$$T_r(T_s(x)) \bmod p \equiv T_{rs}(x) \bmod p \equiv T_s(T_r(x)) \bmod p, \text{ for all } x.$$

Additionally, for any value k and $x \in [-1, 1]$,

$$T_k(x) = cos(k \cdot arccos(x)).$$

Rahasa may wish to use this fact or the recursive nature of Chebyshev polynomials to generate values for large Chebyshev polynomial values.

Recall that $T_{n+1} = 2xT_n(x) - T_{n-1}(x)$, thus

$$\begin{pmatrix} T_1 \\ T_2 \end{pmatrix} = \begin{pmatrix} 0 & 1 \\ -1 & 2x \end{pmatrix} \begin{pmatrix} T_0 \\ T_1 \end{pmatrix}$$

$$\begin{pmatrix} T_n \\ T_{n+1} \end{pmatrix} = \begin{pmatrix} 0 & 1 \\ -1 & 2x \end{pmatrix} \begin{pmatrix} T_{n-1} \\ T_n \end{pmatrix}$$

and more generally,

$$\begin{pmatrix} T_n \\ T_{n+1} \end{pmatrix} = \begin{pmatrix} 0 & 1 \\ -1 & 2x \end{pmatrix}^n \begin{pmatrix} T_0 \\ T_1 \end{pmatrix}.$$

To begin with, Rahasa generates a large prime number s and an integer $x \in [-1, 1]$. For this exercise, Rahasa chooses $s = 2731$ and $x = .065$. Rahasa's public key code is

$$(x, T_s(x)) = (.065, T_{2731}(.065)) = (.065, -0.990181).$$

Siri receives this information and chooses another prime number $r = 1951$ and calculates

$$T_r(x) = T_{1951}(.065) = -0.946032,$$

which is then in turn shared with Siri. Now both Rahasa and Siri can calculate $T_{rs}(x)$,

$$T_{rs}(x) = T_{1951}(-0.990181) = T_{2731}(-0.946032) = 0.950496,$$

even though Rahasa does not know r and Siri does not know s.
From here, Siri would like to determine s with the knowledge that

$$T_s(x) = cos(s \cdot arccos(x)) = -0.990181 \text{ and } arccos(x) = 1.50575.$$

Additionally, Siri knows that

$$T_s(T_r(x)) = cos(s \cdot arccos(T_r(x)) = 0.950496 \text{ with } arccos(T_r(x)) = 2.81156.$$

Thus the possible values for s are

$$s = \left\{ \frac{\pm arccos(T_s(x))}{arccos(x)} + \frac{2\pi k}{arccos(x)} \middle| k \in \mathbb{Z} \right\}$$
$$\approx \{\pm 1.99325 + 2\pi k \cdot 0.664121 | k \in \mathbb{Z}\}.$$

And,

$$s = \left\{ \frac{\pm arccos(T_{rs}(x))}{arccos(T_r(x))} + \frac{2\pi j}{arccos(T_r(x))} \middle| j \in \mathbb{Z} \right\}$$
$$\approx \{\pm 0.112382 + 2\pi j \cdot 0.355674 | j \in \mathbb{Z}\}.$$

Then Siri must determine prime numbers u_1 and u_2 such that

$$1.99325 + 2\pi k_1 \cdot 0.664121 = u_1 \text{ and } -1.99325 + 2\pi k_2 \cdot 0.664121 = u_2$$

$$0.112381 + 2\pi j_1 \cdot 0.355675 = u_1 \text{ and } -0.112381 + 2\pi j_2 \cdot 0.355675 = u_2.$$

With a quick search, Siri knows that there must be integers k_1 and j_2 such that

$$1.99325 + 2\pi k_1 \cdot 0.664121 = -0.112381 + 2\pi j_2 \cdot 0.355675 \text{ is prime,}$$

determining that $k_1 = 654$ and $s = 2731$.

Then if Rahasa sends Siri the entire message,

$$\begin{pmatrix} 51167940 & 941841 & 9 & 49485 & 51167940 \\ 522216 & 39590091 & 51167940 & 39590091 & 51167940 \end{pmatrix}.$$

Rahasa could find the Chebyshev polynomial $T_{2731}(x)$ or $T_5(x) = T_{2731 \bmod 5}(x)$.

$$T_5(x) = 16x^5 - 20x^3 + 5x$$

and solve $T_5(x) = M$ for each encrypted number M in the message. For example,

$$16x^5 - 20x^3 + 5x = 51167940, \text{ when } x \approx 20.$$

Python and R code related to this case study can be found at Github links 26 and 29.

4.8 CASE STUDY: Racial Disparities in Marijuana Arrests

According to ACLU, blacks are 3.6 more likely to be arrested for Marijuana possession than whites in America despite similar usage rates [1]. Since the mid 1980s the United States has put in place aggressive law enforcement to tackle the war on drugs. The United States incarcerates more people than any other nation in the world, largely due to drug possession and usage. Unfortunately, this mass incarceration program, lead by misguided and harsh laws have lead to a systemic problem targeting people of color in America.

We begin by creating a Lagrange interpolation function to fit the Black Arrest Rates as a function of time using the data from Table 4.5. We will assume that year 2010 is year 1.

TABLE 4.5
Black and White Marijuana Arrest Rates per 100K (2010–2018)[1].

Year	Marijuana Possession Arrest Rate	Black Arrest Rate	White Arrest Rate
2010	250.52	659.06	199.19
2011	229.69	624.43	178.43
2012	217.79	601.68	168.75
2013	253.51	625.68	212.55
2014	199.40	552.13	155.80
2015	174.06	459.89	138.90
2016	179.99	477.64	143.42
2017	207.44	560.08	160.60
2018	203.88	567.51	156.06

Matrices, V^*, V_0, V_1, $\ldots$, and V_8 are created.

$$
V^* = \begin{pmatrix}
1 & 1 & 1 & 1 & 1 & 1 & 1 & 1 & 1 \\
1 & 2 & 4 & 8 & 16 & 32 & 64 & 128 & 256 \\
1 & 3 & 9 & 27 & 81 & 243 & 729 & 2187 & 6561 \\
1 & 4 & 16 & 64 & 256 & 1024 & 4096 & 16384 & 65536 \\
1 & 5 & 25 & 125 & 625 & 3125 & 15625 & 78125 & 390625 \\
1 & 6 & 36 & 216 & 1296 & 7776 & 46656 & 279936 & 1679616 \\
1 & 7 & 49 & 343 & 2401 & 16807 & 117649 & 823543 & 5764801 \\
1 & 8 & 64 & 512 & 4096 & 32768 & 262144 & 2097152 & 16777216 \\
1 & 9 & 81 & 729 & 6561 & 59049 & 531441 & 4782969 & 43046721
\end{pmatrix}
$$

$$
V_0 = \begin{pmatrix}
1 & x & x^2 & x^3 & x^4 & x^5 & x^6 & x^7 & x^8 \\
1 & 2 & 4 & 8 & 16 & 32 & 64 & 128 & 256 \\
1 & 3 & 9 & 27 & 81 & 243 & 729 & 2187 & 6561 \\
1 & 4 & 16 & 64 & 256 & 1024 & 4096 & 16384 & 65536 \\
1 & 5 & 25 & 125 & 625 & 3125 & 15625 & 78125 & 390625 \\
1 & 6 & 36 & 216 & 1296 & 7776 & 46656 & 279936 & 1679616 \\
1 & 7 & 49 & 343 & 2401 & 16807 & 117649 & 823543 & 5764801 \\
1 & 8 & 64 & 512 & 4096 & 32768 & 262144 & 2097152 & 16777216 \\
1 & 9 & 81 & 729 & 6561 & 59049 & 531441 & 4782969 & 43046721
\end{pmatrix}
$$

Recall that V_i is created by removing the $i + 1^{\text{st}}$ row of V^* and adding the powers of x as the first row.

$$
P(x) = \sum_{j=0}^{n} (-1)^j y_j \frac{|V_j|}{|V^*|}
$$

$$
= -0.00425025x^8 + 0.257217x^7 - 6.03801x^6 + 72.9773x^5 - 497.339x^4
$$
$$
+ 1944.47x^3 - 4213.56x^2 + 4541.35x - 1183.05.
$$

Perhaps we wish to integrate the Lagrange interpolant that we just created into the model, but we want to minimize Runge's phenomenon. Let's create a sixth-degree Chebyshev interpolant to this data using the function $P(x)$, the Lagrange interpolant, as the function.

We first have to use a linear transformation to map the domain $[1, 9]$ to $[-1, 1]$,

$$u = 4x + 5.$$

Recall that

$$T_n(x) = \begin{pmatrix} 0 & 1 \\ -1 & 2x \end{pmatrix}^n \begin{pmatrix} 1 & x \end{pmatrix}.$$

The roots of $T_7(x) = 128x^8 - 256x^6 + 160x^4 - 32x^2 + 1$ are

$$x_k = \cos\left(\frac{(2k+1)\pi}{16}\right), k = 0, 1, \ldots, 7,$$

$$u_k = \frac{x_k - 5}{4}, k = 0, 1, \ldots, 7,$$

$$c_0 = \frac{1}{8}\sum_{i=0}^{6} P(u_i) \cdot T_0(x_i) \approx 583.06,$$

$$c_k = \frac{2}{8}\sum_{i=0}^{6} P(u_i) \cdot T_k(x_i), k = 1, 2, \ldots, 7.$$

$$\{c_1, c_2, \ldots, c_7\} = \{-66.62, 49.51, 25.94, 2.034, -27.85, -18.711, 22.30\}.$$

$$p(x) = \sum_{i=0}^{7} c_i T_i\left(\frac{x-5}{4}\right)$$
$$= 643.905 - 43.277x + 10.938x^2 + 8.1211x^3 - 4.8833x^4 + 0.7772x^5$$
$$- 0.0383x^6$$

What questions would you start to ask in order to determine racial inequity related to marijuana arrests?

Figure 4.11 shows the yearly black arrest rates centered with bands of width 3 standard deviations as well as the yearly white arrest rates centered with bands of width 3 standard deviations relative to that data set. This shows that the two data set are extremely far from one another. One might consider following up this case study with a statistical hypothesis test.

Sample Python and R code for this case study can be found at Github links 27 and 30.

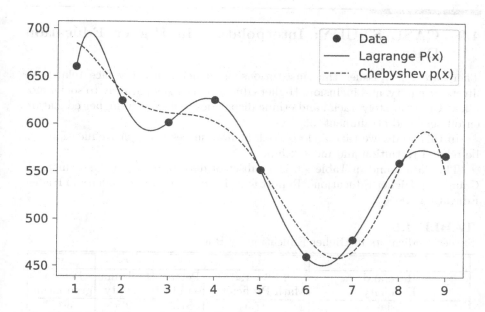

FIGURE 4.10
Lagrange interpolant, $P(x)$, and Chebyshev interpolant, $p(x)$, fitted to the data
from Table 4.5.

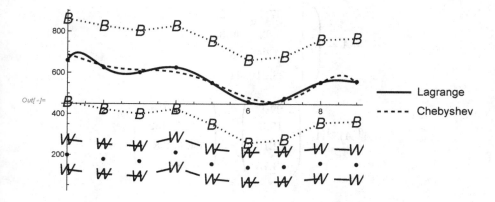

FIGURE 4.11
Data $\pm 3\sigma$ bands.

4.9 CASE STUDY: Interpolation in Higher Education Data

There are many issues that institutions of higher education face related to diversity, equity, and inclusion. Higher education is a key pathway to social mobility. Unfortunately, racial and ethnic disparities still exist in higher education enrollment and attainment [33].

In this study, we take a closer look at how success in higher education can be related to tuition and mentorship.

The data found in Table 4.6 is a subset of data provided by the American Council on Higher Education's Report focusing on Race and Ethnicity in Higher Education [18].

TABLE 4.6
Success Indicators in Higher Education by Race.

	x_1	x_2	x_3	y
Race and Ethnicity	12th Grade Math Proficiency	Percent enrolled	Full time faculty	Percent graduating
Asian	37.5	57.21	9.5	30.7
Black or African American	6.83	36.33	5.7	15.3
Hispanic or Latino	11.25	38.32	4.8	12.2
White	28.21	41.86	72.6	23.75

In this study, we will look at how factors from Table 4.6 can be used to predict graduation rate and future success.

Let

$$A = \begin{pmatrix} 1 & 37.5 & 57.21 & 9.5 \\ 1 & 6.83 & 36.33 & 5.7 \\ 1 & 11.25 & 38.32 & 4.8 \\ 1 & 28.21 & 41.86 & 72.6 \end{pmatrix}$$

$$(A^T A)^{-1} A^T \begin{pmatrix} 30.7 \\ 15.3 \\ 12.16 \\ 23.75 \end{pmatrix} = \begin{pmatrix} -131.767 \\ -2.59432 \\ 4.44338 \\ 0.588323 \end{pmatrix}$$

and

$$\widehat{gradrate} = -131.767 - 2.59432x_1 + 4.44338x_2 + 0.588323x_3$$
$$SSE = 9.4\dot{1}0^{-23}.$$

Graduation is not always the only measure of success. A measure of success after graduation may take into account debt and the ability to pursue a career that gives a student the ability to pay back the debt from achieving their degree.

TABLE 4.7
Data related to borrowing for college.

x_4	y_2
Percent borrowing to achieve degree	Ratio of Amount Still Owed to Amount Borrowed
6.6	45.6
12	105.5
15	69.7
58.5	53.5

We wish to create a 3rd degree Hermitian interpolant, $h(x)$, using the percent borrowed to predict the ratio still owed to the amount borrowed, but we do not know any information about $h'(x)$.

One strategy is to estimate derivatives with slopes of secant lines between consecutive points. This strategy does make some assumptions and does a better job when consecutive points are close together. We will also assume that the derivative remains the same for the last two data points.

TABLE 4.8
Estimated $h'(x)$.

x_4	Approximate $h'(x_4)$
6.6	$\frac{105.5-45.6}{12-6.6} = 11.0926$
12	$\frac{69.7-105.5}{15-12} = -11.9333$
15	$\frac{53.5-69.7}{58.5-15} = -0.372414$
58.5	

Here we employ the divided difference method.

$$f[z_0, z_1, z_2, z_3] = \frac{0.248621 + 2.7411817}{58.5 - 6.6}.$$

TABLE 4.9

Divided difference table.

z	$f[z]$	$f[z_i, z_{i+1}]$	$f[z_i, z_{i+1}, z_{i+2}]$
$z_0 = 6.6$	45.6		
		$f[z_0, z_1] = 11.0926$	
$z_1 = 12$	105.5		
			$f[z_0, z_1, z_2] = \frac{-11.9333 - 11.0926}{15 - 6.6}$ $= -2.7411817$
		$f[z_1, z_2] = -11.9333$	
$z_2 = 15$	69.7		
			$f[z_1, z_2, z_3] = \frac{-0.372414 + 11.9333}{58.5 - 12}$ $= 0.248621$
		$f[z_2, z_3] = -0.372414$	
$z_3 = 58.5$	53.5		

Then,

$$h(x) = 45.6 + 11.0926(x - 6.6) - 2.7412(x - 6.6)(x - 12)$$
$$+ 0.0576(x - 6.6)(x - 12)(x - 15)$$
$$= 0.0576x^3 - 4.67656x^2 + 82.7112x - 313.143$$

This small data set found in Table 4.10, was constructed using five east coast schools identified as top institutions awarding degrees to minority students [22] and seven additional east coast schools chosen at random.

The variables included in Table 4.10 are in state tuition, total enrollment, minority enrollment, and graduation rates for the white and black (nonhispanic) populations at each of the institutions.

We begin by exploring models that use the tuition and percent of enrolled minority students to predict the graduation rate of black students at an institution, using the data from Table 4.10.

Using a least squares regression model, with $\hat{y} = graduation\ rate$,

$$\hat{y} = 0.64419 + 1.997644 \cdot 10^{-6} (tuition)$$
$$- 0.036985(\%\ of\ enrolled\ minority\ students).$$

We can also normalize this data and use principal component analysis to see which schools in Table 4.10 are similar. Python and R code for this case study can be found at Github links 28 and 31.

TABLE 4.10
University and College Diversity Data.

	College	InState Tuition	Total Enrollment	Total Minority	Graduation Rate for Black Students in 2020
1	Carnegie Mellon	69883	12587	3392	80
2	Clemson	25802	21857	2849	63.4
3	Florida State	17332	41226	12238	72.5
4	James Madison	22108	20855	3574	74.1
5	High Point	49248	4399	668	74.3
6	New England College	52136	2399	638	19.2
7	Old Dominion	22492	24932	10059	53.14
8	Providence College	65090	4533	652	84.8
9	Rutgers	27680	48378	21656	79.86
10	U. of CT	28604	26541	6255	71.05
11	U. of Penn	71200	24806	7805	94.22
12	William and Mary	35636	8437	2135	88.18

4.10 Exercises

1. Table 4.11 shows some racial demographics related to Covid-19 deaths, as of February 2021, by state.

   a. Use least squares regression to fit a 4th degree polynomial to the data represented by % of the population that is white versus the % of cases from the white population.

   b. Determine the SSE related to the model in part a.

   c. Relative error can be a better expression of error relate to the size of the data. The relative l_2 error is

   $$\frac{\sqrt{\sum(y - \hat{y})^2}}{\sqrt{\sum y^2}}.$$

   Determine the relative l_2 error for the model in part a.

TABLE 4.11
Percent of Covid-19 cases by race.

State	% of Population that is white	% of Cases that are from the white population	% of Population that is black	% of Cases that are from the black population
CA	.36	.20	.05	.04
FL	.53	.40	.15	.15
PA	.80	.82	.11	.14
TX	.41	.36	.12	.19
WA	.48	.68	.04	.06

2. Use Table 4.11 and Vandermonde determinants to create a 4th degree Lagrange interpolation polynomial that matches the points represented by % of the population that is black versus the % of cases from the black population.

3. Given the data in Table 4.11, and an additional data point $\{0, 0\}$, use the Vandermonde determinants to create a 5th degree Lagrange interpolation polynomial that matches the points described in Problem 2.

4. Determine the 4th- and 5th-Chebyshev polynomials of the first kind and show that they are orthogonal under the inner product defined in this chapter for the set of Chebyshev polynomials.

5. The nth-degree Chebyshev polynomial of the second kind is defined as the determinant of $n \times n$ matrix of the form

$$\begin{pmatrix} x & 1 & 0 & \cdots & \cdots & \cdots & 0 \\ 1 & x & 1 & 0 & \cdots & \cdots & 0 \\ 0 & 1 & x & 1 & 0 & \cdots & 0 \\ \vdots & \ddots & \ddots & \ddots & \vdots & \vdots & \vdots \\ 0 & 0 & \cdots & 0 & 1 & x & 1 \\ 0 & 0 & \cdots & \cdots & 0 & 1 & x \end{pmatrix}.$$

Determine the 3rd- and 4th-degree Chebyshev polynomials of the second kind and determine if they are orthogonal under the inner product defined in this chapter for the set of Chebyshev polynomials.

6. Show that Hermitian polynomials $H_2(x)$ and $H_3(x)$ are orthogonal functions.

7. Find the LU-decomposition for

$$\begin{pmatrix} 1 & 1 & 2 \\ 1 & 2 & 5 \\ 1 & -1 & 3 \end{pmatrix}.$$

8. Table 4.3 shows data for the growth of COVID 19 over a period of months,

   a. Create the Vandermonde matrix with the data from Table 4.3.
   b. Find the LU-decomposition of the Vandermonde matrix in part a.
   c. Find $f[x_0]$, $f[x_0, x_1]$, $f[x_0, x_1, x_2]$, and $f[x_0, x_1, x_2, x_3]$ using the LU-decomposition from part b.

d. Use the divided differences from part c. to find a fifth degree Newton's interpolation polynomial to fit the data.

9. In addition to the data presented in Table 4.3, the instantaneous growth rate is given in Table 4.12. Use this information to construct a Hermite interpolant that matches both the data from Table 4.3 and Table 4.12.

TABLE 4.12
Growth rate of COVID 19 total cases in NC in 2020.

Date	July 1	Aug. 1	Sept. 1	Oct. 1	Nov. 1	Dec 1
Growth rate	.9	.4	.25	.3	.35	.5

10. One in 7 women and 1 in 25 men have been injured by an intimate partner [24]. According to Evans et al. [10], with personal mobility limited by COVID-19 stay at home orders, advocates expressed concern about a potential increase in intimate partner violence (IPV); however, call centers experienced up to a 50% decrease in calls in some regions [10]. In this problem, we investigate the effects of COVID-19 on reported cases of IPV.

TABLE 4.13
Number of national contacts, calls, chats, and emails from concerned parties, in the first half of each year 2017–2020 (National Domestic Violence Hotline [23]).

Year	2016	2017	2018	2019	2020
National contacts	124712	106276	98954	156429	152415

a. Create a least squares third degree polynomial interpolant for the data provided in Table 4.13.

b. Table 4.14 shows the rate of change of calls to the National Domestic Hotline from 2017–2020. Use this information together with the data in Table 4.14 to create a Hermitian polynomial.

TABLE 4.14
Rate of change of calls to the National Domestic Hotline from 2017–2020.

Year	2017	2018	2019	2020
Percent increase	-.2	-.1	.6	-.1

c. Use your polynomial from part b. to estimate the number of calls to the National Domestic Hotline midway through the year in 2021.

d. While making an estimate within the domain of the given data is called interpolation, making estimates outside of the domain of the given data is called **extrapolation**. Use the interpolant from part b. to estimate the number of calls in 2022.

e. Discuss what issues can arise from estimation through extrapolation.

5

Optimization and Learning Techniques for Regression

In this chapter, we look at algorithms that combine topics that we have touched on in previous chapters with machine learning. A common goal of many of these algorithms is the importance of optimizing, either minimizing or maximizing, a chosen function. These ideas require some basic knowledge of probability theory, differentiation, and integration, and thus we start with an overview of these topics.

5.1 Basics of Probability Theory

A **probability space** is defined by a sample space, S, a set of events, E, and a probability measure $P : E \rightarrow [0,1]$ where $P(S) = 1$.

A **random variable**, X, is a function on the sample space. A **discrete random variable** is a random variable whose sample space is countable, finite, or infinite. A **continuous random variable** is one in which the sample space is uncountable.

If X is a discrete random variable,

$$P(X) = P(X = x)$$

is called a **probability distribution function**.

For a discrete random variable, X, the **cumulative distribution function**,

$$D(x) = P(X \leq x_k) = \sum_{i=1}^{k} P(X = x_i),$$

where the sample space $S = \{x_1, x_2, \ldots, x_k, \ldots, x_n\}$.

The **expected value**, or mean, of a discrete random variable, X,

$$E(X) = \sum_{i=1}^{n} x_i \cdot P(X = x_i),$$

DOI: 10.1201/9781003025672-5

and the variance of a discrete random variable,

$$Var(X) = E(X^2) - (E(X))^2 = \sum_{i=1}^{n} x_i^2 \cdot P(X = x_i) - \left(\sum_{i=1}^{n} x_i \cdot P(X = x_i) \right)^2 .$$

Example 5.1. *Consider an experiment where a fair coin is tossed exactly 10 times and the random variable, X, the number of heads, is recorded.*

*The sample space $S = \{0, 1, 2, 3, 4, 5, 6, 7, 8, 9, 10\}$ has a finite number of elements and thus this is an example of a **finite probability space**.*

*The random variable, X, in this example is a discrete random variable and the probability distribution function is an example of a **Binomial Distribution**.*

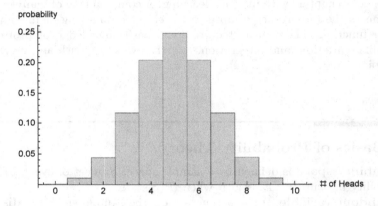

FIGURE 5.1
Binomial Distribution representing the number of heads in 10 flips of a fair coin.

The binomial distribution representing the probability that $P(X = x)$, where x is the number of heads, is shown in Figure 5.1.

In general, if a random variable, x, has a binomial probability distribution, denoted $x \sim Bi(n,p)$ then

$$P(X = x) = \begin{cases} \binom{n}{x} p^x (1 - p)^{n-x}, & \text{if } x \in \{1, 2, \ldots, n\}, \\ 0, & \text{otherwise,} \end{cases}$$

$$E(x) = np,$$
$$Var(x) = np(1 - p),$$

where p represents the probability of success in the experiment and n represents the number of trials of the experiment.

Another discrete probability distribution of interest is the **Poisson Distribution**.

The Poisson distribution is a special type of Binomial Distribution that focuses on the average time between successes.

If the random variable X is the number of successes in a given time period and if λ is the average time between successes then X has a Poisson distribution, $X \sim Pois(\lambda)$, with $E(X) = \lambda$ and $Var(X) = \lambda$,

$$P(X = x) = \frac{\lambda^x e^{-\lambda}}{x!}.$$

Example 5.2. *Acme industries just released a new website. In the first day after the release, they see that they are averaging 10 hits on the site per hour. Using this information, and assuming that the average is $\lambda = 10$ hits on the site per hour, the probability that on the next day they would experience 15 hits in the first hour is*

$$P(X = 10 \ hits) = \frac{10^{15}}{15!}e^{-10} \approx 0.03472.$$

and the probability that they will have more than 15 hits in the first hour is

$$P(X > 15) = 1 - P(X \le 15) = \sum_{i=0}^{15} \frac{10^i}{i!}e^{-10} \approx 0.04874.$$

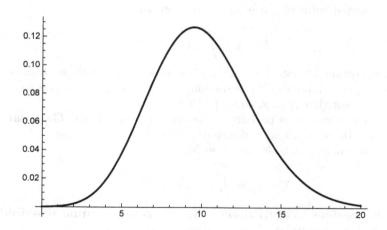

FIGURE 5.2
Poisson Distribution with $\lambda = 10$.

When you compare the shape of the curve in Figures 5.1 and 5.2, you might notice that they are very similar.

If in fact we let $\lambda = np$,

$$lim_{n \to \infty} P(X = k) = lim_{n \to \infty} \binom{n}{k} p^k (1-p)^{n-k}$$

$$= lim_{n \to \infty} \frac{n!}{(k!)(n-k)!} \left(\frac{\lambda}{n}\right)^k \left(1 - \frac{\lambda}{n}\right)^{n-k}$$

$$= \left(\frac{\lambda}{k!}\right)^k lim_{n \to \infty} \frac{n!}{n^k (n-k)!}$$

$$= \left(\frac{\lambda}{k!}\right)^k e^k$$

We will also be looking at problems that involve **infinite probability spaces** and have continuous probability measures, called **probability density functions**.

If X is a continuous random variable with probability density function $P(X)$ then $P(X)$ is the derivative of the cumulative distribution function, $D(x)$, where

$$D(x) = P(X \le x) = \int_{-\infty}^{x} P(t)dt.$$

The expected value of a continuous random variable, X,

$$E(X) = \int_{-\infty}^{x} t \cdot P(t)dt.$$

It is important to note that for discrete random variable, we focus on the chance of $X = x$; however, if the random variable X is continuous we do not discuss the event that $X = x$, since $\int_{x}^{x} P(t)\, dt = 0$.

A common continuous probability density functions is the **Gaussian distribution** with mean μ and variance σ^2, $N(\mu, \sigma^2)$.

If X is a random variable and $X \sim N(\mu, \sigma^2)$, then

$$P(X \le x) = \int_{-\infty}^{x} \frac{1}{\sqrt{2\pi\sigma^2}} e^{\frac{-(t-\mu)^2}{2\sigma^2}}\, dt.$$

Another continuous distribution of interest is the **Gamma distribution**. Figure 5.3 shows examples of each of these distributions.

Many times we will also be looking at how multiple random variables interact and effect each others' behavior.

If X and Y are two discrete random variables, the **joint distribution** of X and Y, denoted $f_{XY}(x,y)$, is

$$f_{XY}(x,y) = P(X = x \cap Y = y).$$

If X and Y are continuous random variables, then

$$f_{XY}(x,y) = P(X \le x \cap Y \le y).$$

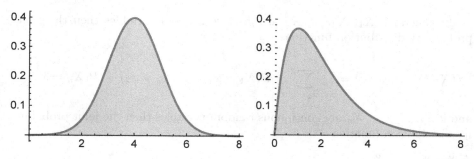

FIGURE 5.3
(Left) Gaussian distribution with parameters $\mu = 4$, $\sigma^2 = 1$ and (Right) Gamma distribution with parameters $\alpha = 2$, $\beta = 1$.

One can also look at **conditional probability distributions** where we explore the probability of a random variable Y occurring given that a random variable X has occurred.

$$f_{Y|X}(y|x) = P(Y = y|X = x) = \frac{P(Y = y \cap X = x)}{P(X = x)}$$

for discrete random variables or for continuous random variables,

$$f_{Y|X}(y|x) = P(Y \leq y|X \leq x) = \frac{P(Y \leq y \cap X \leq x)}{P(X \leq x)}.$$

Two random variables, X and Y, are **independent** if

$$P(X = x \cap Y = y) = P(X = x)P(Y = y)$$

and thus $P(Y = y|X = x) = P(Y|y)$.

Example 5.3. *If X is the random variable representing the number of heads that occur in 8 tosses of a fair coin and Y is the random variable representing the number of heads in 10 tosses of a fair coin.*

$$P(Y = y|X = 4) = \begin{cases} \frac{1}{4}, & \text{if } y = 4, \\ \frac{1}{2}, & \text{if } y = 5, \\ \frac{1}{4}, & \text{if } y = 6. \end{cases}$$

Notice that $P(Y = y)$ would be the same distribution as that for the random variable represented in Example 5.1, which is not the same as $P(Y = y|X = 4)$ or $P(Y = y|X = x)$ for any value of $x = 0, 1, 2, \ldots, 10$.

A set of random variables is called **independent and identical distributed (i.i.d.)** if each of the random variables in the set has the same probability distribution and they are mutually independent.

In general if X_1, X_2, ..., X_n are discrete random variables then the joint probability distribution function

$$f(X_1, X_2, \ldots, X_n) = \sum_{x_1, \, x_2, \ldots, \, x_n} P(X_1 = x_1) \cdot P(X_2 = x_2) \cdots P(X_n = x_n),$$

and if X_1, X_2, $\ldots$, X_n are continuous random variables then the joint probability distribution function

$$f(X_1, X_2, \ldots, X_n)$$
$$= \int_{x_1, \, x_2, \ldots, \, x_n} P(X_1 = x_1) \cdot P(X_2 = x_2) \cdots P(X_n = x_n) dX_1 dX_2 \ldots dX_n.$$

Example 5.4. *Suppose, in an experiment, a fair coin is flipped three times. Define the random variable X as the random variable representing the number of heads that occurred, and Y as the random variable representing the number of heads in the first two tosses.*

The variables in this example are dependent and are not identically distributed. $f(X,Y)$ can be seen in Table 5.1.

TABLE 5.1
Example of a discrete joint
probability distribution.

		Y		
		0	1	2
	0	$\frac{1}{2}$	0	0
X	1	$\frac{1}{2}$	$\frac{1}{2}$	0
	2	0	$\frac{1}{2}$	$\frac{1}{2}$
	3	0	0	$\frac{1}{2}$

Example 5.5. *Given two i.i.d. random variables, $X, Y \sim N(\mu,\sigma^2)$. The joint distribution*

$$f_{XY}(x,y) = \frac{1}{2\pi\sigma^2} e^{-((x-\mu)^2 + (y-\mu)^2)/2\sigma^2}.$$

Theorem 13. *If X_1, $X_2,\ldots$, X_n are i.i.d. random variables with probability distribution functions, $f_{X_i}(x_i)$ respectively, expected value, or mean, of X_i, $E(X_i) = \mu$ and variance of X_i, $Var(X_i) = \sigma^2$ then their joint probability distribution function*

$$f_{X_1 \ldots X_n}(x_1 \ldots x_n) = f_{X_1}(x_1) \ldots f_{X_n}(x_n),$$
$$E(X_1 + X_2 + \cdots + X_n) = n\mu \text{ and}$$
$$Var(X_1 + X_2 + \cdots + X_n) = n\sigma^2.$$

Now that we have introduced the basics of probability theory, we are ready to introduce you to some matrix calculus. Both probability theory and matrix calculus will be applied in Chapters 5 and 7.

5.2 Introduction to Matrix Calculus

As we saw in the previous section, there are times in which integration must be employed in order to determine the cumulative distribution function from a probability density function. Additionally, since we are focusing on optimization in this chapter, it is important to understand certain topics related to differentiation with a focus toward how to apply these concept to matrices.

5.2.1 Matrix Differentiation

Given a scalar vector, $X = \begin{pmatrix} x_1 \\ x_2 \\ \vdots \\ x_n \end{pmatrix}$ and function vector

$$Y = \begin{pmatrix} f_1(x_1,x_2,\ldots,x_n) \\ f_2(x_1,x_2,\ldots,x_n) \\ \vdots \\ f_m(x_1,x_2,\ldots,x_n) \end{pmatrix},$$

the derivative of Y with respect to a scalar, x_i, $1 \leq i \leq n$, is

$$\frac{\partial Y}{\partial x_i} = \left(\frac{\partial f_1}{\partial x_i}, \frac{\partial f_2}{\partial x_i}, \ldots, \frac{\partial f_m}{\partial x_i} \right)$$

and the derivative of Y with respect to X is a matrix

$$\frac{\partial Y}{\partial X} = \begin{pmatrix} \frac{\partial f_1}{\partial x_1} & \frac{\partial f_1}{\partial x_2} & \cdots & \frac{\partial f_1}{\partial x_n} \\ \frac{\partial f_2}{\partial x_1} & \frac{\partial f_2}{\partial x_2} & \cdots & \frac{\partial f_2}{\partial x_n} \\ \vdots & \ddots & \ddots & \vdots \\ \frac{\partial f_m}{\partial x_1} & \frac{\partial f_m}{\partial x_2} & \cdots & \frac{\partial f_m}{\partial x_n} \end{pmatrix}.$$

Given a transformation $Y = Y(X)$, where X and Y are of the same dimension, the **Jacobian** is

$$|J| = \frac{\partial Y}{\partial X}$$

Example 5.6. *If* $Y(X) = a_0 + a_1 X$ *with* $Y_i(x_i) = x_i$, $X = (x_1, \ldots, x_n)$ *and* a_0 *and* a_1 *are constants then*

$$
|J| = \begin{vmatrix} \frac{\partial(a_0 + a_1 x_1)}{\partial x_1} & 0 & \cdots & 0 \\ 0 & \frac{\partial(a_0 + a_1 x_2)}{\partial x_2} & 0 & \cdots \\ \vdots & & \ddots & \ddots & \vdots \\ 0 & 0 & \cdots & \frac{\partial(a_0 + a_1 x_n)}{\partial x_n} \end{vmatrix}
$$

$$
= \begin{vmatrix} a_1 & 0 & \cdots & 0 \\ 0 & a_1 & 0 & \cdots \\ \vdots & \ddots & \ddots & \vdots \\ 0 & 0 & \cdots & a_1 \end{vmatrix} = (a_1)^n.
$$

Example 5.7. *If* $Y = \begin{pmatrix} r\cos(\theta) & r\sin(\theta) \end{pmatrix}$ *and* $X = \begin{pmatrix} r & \theta \end{pmatrix}$ *then*

$$
\frac{\partial Y}{\partial r} = (\cos(\theta), \sin(\theta))
$$

and

$$
\frac{\partial Y}{\partial X} = \begin{pmatrix} \cos(\theta) & -r\sin(\theta) \\ \sin(\theta) & r\cos(\theta) \end{pmatrix}.
$$

In addition,

$$
|J| = (r\cos(\theta))^2 + (r\sin(\theta))^2 = r.
$$

If instead, $y = f(X)$ is a scalar function of matrix

$$
X = \begin{pmatrix} x_{11} & x_{12} & \cdots & x_{1n} \\ x_{21} & x_{22} & \cdots & x_{2n} \\ \vdots & \ddots & \ddots & \vdots \\ x_{m1} & x_{m2} & \cdots & x_{mn} \end{pmatrix}, \text{ then}
$$

$$
\nabla y = \frac{\partial y}{\partial X} = \begin{pmatrix} \frac{\partial y}{\partial x_{11}} & \frac{\partial y}{\partial x_{12}} & \cdots & \frac{\partial y}{\partial x_{1n}} \\ \frac{\partial y}{\partial x_{21}} & \frac{\partial y}{\partial x_{22}} & \cdots & \frac{\partial y}{\partial x_{2n}} \\ \vdots & \ddots & \ddots & \vdots \\ \frac{\partial y}{\partial x_{m1}} & \frac{\partial y}{\partial x_{m2}} & \cdots & \frac{\partial y}{\partial x_{mn}} \end{pmatrix}
$$

$$
= \sum_{i,j} E \frac{\partial y}{\partial x_{ij}},
$$

where $E_{i,j} = 1$ and all other entries are zero and ∇y is called the **gradient matrix**.

Example 5.8. *If $y = tr(X)$ then $\nabla y = I$. If X is a 2×2 matrix and*

$$y = |X| = x_{11}x_{22} - x_{12}x_{21}$$

then

$$\nabla y = \begin{pmatrix} x_{22} & -x_{21} \\ -x_{12} & x_{11} \end{pmatrix}.$$

The **Hessian matrix**, H_y, is a square matrix of second order partial derivatives.

$$H_y = \begin{pmatrix} \frac{\partial^2 y}{\partial x_{11}^2} & \frac{\partial^2 y}{\partial x_{11}\partial x_{12}} & \cdots & \frac{\partial^2 y}{\partial x_{11}\partial x_{1n}} \\ \frac{\partial^2 y}{\partial x_{22}\partial x_{21}} & \frac{\partial^2 y}{\partial x_{22}^2} & \cdots & \frac{\partial^2 y}{\partial x_{22}\partial x_{2n}} \\ \vdots & \ddots & \ddots & \vdots \\ \frac{\partial^2 y}{\partial x_{nn}\partial x_{n1}} & \frac{\partial^2 y}{\partial x_{nn}\partial x_{n2}} & \cdots & \frac{\partial^2 y}{\partial x_{nn}^2} \end{pmatrix}.$$

Example 5.9. *If $f(x_1,x_2,x_3) = x_1^2 + x_2^2 + x_3^2$ then*

$$\nabla f = \begin{pmatrix} 2x_1 & 0 & 0 \\ 0 & 2x_2 & 0 \\ 0 & 0 & 2x_3 \end{pmatrix},$$

and the Hessian matrix

$$H_f = \begin{pmatrix} 2 & 0 & 0 \\ 0 & 2 & 0 \\ 0 & 0 & 2 \end{pmatrix}.$$

Example 5.10. *If $X = \begin{pmatrix} x_{11} & x_{12} \\ x_{21} & x_{22} \end{pmatrix}$ and $y = e^{|X|} = e^{x_{11}x_{22} - x_{12}x_{21}}$, then*

$$\nabla y = \begin{pmatrix} x_{22}(e^{x_{11}x_{22} - x_{12}x_{21}}) & -x_{21}(e^{x_{11}x_{22} - x_{12}x_{21}}) \\ -x_{12}(e^{x_{11}x_{22} - x_{12}x_{21}}) & x_{11}(e^{x_{11}x_{22} - x_{12}x_{21}}) \end{pmatrix}$$

$$= y|X|X^{-T}.$$

The Hessian matrix

$$H_y = \begin{pmatrix} x_{11}x_{22}(e^{x_{11}x_{22}-x_{12}x_{21}}+1) & -x_{21}x_{22}(e^{x_{11}x_{22}-x_{12}x_{21}}) \\ -x_{11}x_{12}(e^{x_{11}x_{22}-x_{12}x_{21}}) & x_{11}x_{22}(e^{x_{11}x_{22}-x_{12}x_{21}}+1) \end{pmatrix}$$

$$= y|X|X^{-T} + \begin{vmatrix} x_{11} & 0 \\ 0 & x_{22} \end{vmatrix} I.$$

If $y = f(X)$ is a scalar function where $X = (x_1, \ldots, x_n)$, we define the **ordinary differential** as

$$df = \sum_{i=1}^{n} \frac{\partial f}{\partial x_i} dx_i.$$

Whereas, if $X = [x_{i,j}]$ is an $m \times n$ matrix, then we define the **matrix differential** as

$$dX = \begin{pmatrix} dx_{11} & dx_{12} & \cdots & dx_{1n} \\ dx_{21} & dx_{22} & \cdots & dx_{2n} \\ \vdots & \ddots & \ddots & \vdots \\ dx_{m1} & dx_{m2} & \cdots & dx_{mn} \end{pmatrix}.$$

The matrix differential satisfies the properties of a linear operator,

$$d(kX) = kdX, \text{ and } d(X+Y) = dX + dY,$$

where k is a scalar and additionally satisfies the chain rule

$$d(XY) = (d(X))Y + Xd(Y),$$

if X and Y are product conforming matrices.

Example 5.11. *Let X be a 2×2 matrix and define*

$$f(X) = |X|, \ g(X) = X^{-1}, \text{ and } h(X) = e^{-X^2/2}.$$

Then,

$$df = d|X| = \sum_{i,j} \frac{\partial |X|}{\partial x_{ij}} dx_{ij}$$

$$= x_{22}dx_{11} - x_{21}dx_{12} - x_{12}dx_{21} + x_{11}dx_{22}$$

$$= tr(|X|X^{-1}dX)$$

Using the fact that $X^{-1}X = I$, $d(X^{-1})X + X^{-1}dX = 0$. *Thus,*

$$dg = d(X^{-1}) = -X^{-1}dX(X^{-1}).$$

We also note that $ln(h(X)) = \frac{-X^2}{2}$. *Thus* $dh(h(x))^{-1} = -XdX$ *and*

$$dh = d(e^{-X^2/2}) = -XdXe^{-X^2/2}.$$

5.2.2 Matrix Integration

If A is a matrix of scalar functions and one wishes to integrate with respect to a scalar variable, x, then

$$\int A \, dx = \int a_{ij} dx,$$

however, if $f(x)$ is a vector function then there are instances when one might want to perform a transformation $x = g(y)$, then

$$\int f(x) dx = \int f(g(y)) \cdot |det(J(y))| dy,$$

where $|det(J(y))|$ is the absolute value of determinant of the Jacobian associated with the transformation $g(y)$.

Example 5.12. *Let* $A = \begin{pmatrix} \sin(t) & \cos(2t) \\ 3t^2 & 1 \end{pmatrix}$,

$$\int_0^\pi A \, dt = \begin{pmatrix} 2 & 0 \\ \pi^3 & \pi \end{pmatrix}.$$

If $f(X)$ is a matrix function and we wish to perform a matrix transformation $Y = g(X)$, then

$$\int f(X) dx = \int f(g^{-1}(Y))|det(J)|dY.$$

Another matrix of importance in the integration of matrices is the square **Vandermonde matrix**.

The Vandermonde matrix plays an important role in signal processing and distribution theory. In this chapter, this is applicable as we wish to find proba- bility distributions and expected values for matrices.

For example, we may wish to determine the following integral, which is related to finding the joint probability distribution of normal distributions with mean μ and variance σ^2, $X_i \sim N(\mu, \sigma^2)$, $i = 1, 2, \ldots, n$,

$$\int_{-\infty}^\infty \cdots \int_{-\infty}^\infty \int_{-\infty}^\infty e^{-(X_1-\mu)^T \Sigma^{-1}(X_1-\mu)/2} \, dX_1 dX_2 \ldots dX_n,$$

where

$$\Sigma = \begin{pmatrix} \sigma_1^2 & cov(X_1,X_2) & cov(X_1,X_3) & \cdots & cov(X_1,X_n) \\ cov(X_1,X_2) & \sigma_2^2 & cov(X_2,X_3) & \cdots & cov(X_2,X_n) \\ \vdots & \ddots & \ddots & \ddots & \vdots \\ cov(X_1,X_n) & cov(X_2,X_n) & \cdots & cov(X_{n-1},X_n) & \sigma_n^2 \end{pmatrix}.$$

We may make the transformation $y_{ij} = x_{i,j} - \mu_{ij}$, then the integral becomes

$$\int_{-\infty}^{\infty} \cdots \int_{-\infty}^{\infty} \int_{-\infty}^{\infty} e^{-(Y)^T \Sigma^{-1}(Y)/2} dY,$$

however, if you make the transformation $Z = P^{-1}Y$ where $P^T \Sigma^{-1} P = I$, then

$$|det(J)| = |det(P)| = |\Sigma|^{1/2}$$

and the integral then becomes

$$\int_{-\infty}^{\infty} \cdots \int_{-\infty}^{\infty} \int_{-\infty}^{\infty} e^{-(PZ)^T \Sigma^{-1}(PZ)/2} |\Sigma|^{1/2} dZ$$

$$= \int_{-\infty}^{\infty} \cdots \int_{-\infty}^{\infty} \int_{-\infty}^{\infty} e^{-Z^T Z/2} |\Sigma|^{1/2} dZ$$

$$= (2\pi)^{n^2/2} |\Sigma|^{1/2}.$$

Using the same transformation,

$$(2\pi)^{-mn/2} |\Sigma|^{-1/2} \int_{\mathbf{R}^{mn}} X e^{-(X-\mu)^T \Sigma^{-1}(X-\mu)/2} dX = \mu.$$

5.3   Maximum Likelihood Estimation

In Chapter 4, many interpolation techniques are presented including Least Squares Regression. In this section, we take another look at least squares regression through a probabilistic lens.

Unlike many of the techniques presented which look to minimize a loss or cost function, the maximum likelihood technique for regression attempts to determine the interpolant with the maximum probability (or likelihood) of occurring.

We begin by assuming the interpolant is a linear function $\hat{y} = a_0 + a_1 x$. We wish to determine the likelihood of this function occurring, $L(a_0, a_1)$,

$$L(a_0, a_1) = \prod_{i=1}^{n} p(y_i | x_i, a_0, a_1),$$

where $p(y_i|\ x_i, a_0, a_1)$ represents the conditional probability that $Y_i = y_i$ given that $X_i = x_i$ and the assumed interpolant $\hat{y} = a_0 + a_1 x$.

Assume that $Y_i \sim N(\mu, s^2)$ are identically independent random variables. That is, for each y_i, where $i = 1, \ldots, n$,

$$p(y_i|\ x_i, a_0, a_1) = \frac{1}{\sqrt{2\pi s^2}} e^{-\frac{(y_i - (a_0 + a_1 x_i))^2}{2s^2}}.$$

Recall that since $Y_1, Y_2, \ldots, Y_n$ are i.i.d. random variables, their joint probability distribution, which we will call $L(a_0, a_1)$ is

$$\prod_{i=1}^{n} p(y_i|\ x_i, a_0, a_1).$$

Before we take the derivative of $L(a_0, a_1)$, in order to find values for a_0 and a_1 that maximize $L(a_0, a_1)$, it is important to note that it is often more convenient to work with the **log-likelihood** function, $LL(a_0, a_1)$, because you then end up taking derivatives of sums of functions rather than products of functions.

Recall that extremal values, maximums and minimums, can occur at critical points, where the derivative of a function is equal to 0.

$$
\begin{aligned}
LL(a_0, a_1) &= \log\left(\prod_{i=1}^{n} p(y_i|\ x_i, a_0, a_1)\right) \\
&= \sum_{i=1}^{n} \log(p(y_i|\ x_i, a_0, a_1)) \\
&= \sum_{i=1}^{n} \log\left(\frac{1}{\sqrt{2\pi s^2}} e^{-\frac{(y_i - (a_0 + a_1 x_i))^2}{2s^2}}\right) \\
&= n \log\left(\frac{1}{\sqrt{2\pi s^2}}\right) - \frac{1}{2s^2} \sum_{i=1}^{n} (y_i - (a_0 + a_1 x_i))^2 \\
&= -\frac{n \log(2\pi)}{2} - 2n \log(s) - \frac{1}{2s^2} \sum_{i=1}^{n} (y_i - (a_0 + a_1 x_i))^2.
\end{aligned}
$$

Determining values, a_0 and a_1, that maximize $LL(a_0, a_1)$ is equivalent to determining values, a_0 and a_1, that maximize $L(a_0, a_1)$ since $log(x)$ is an increasing function.

$$\frac{\partial LL(a_0, a_1)}{\partial a_1} = -\frac{1}{s^2} \sum_{i=1}^{n} x_i(y_i - (a_0 + a_1 x_i)) = 0$$

when

$$a_1 = \frac{\sum_{i=1}^{n} x_i y_i}{\sum_{i=1}^{n} x_i}$$

$$= \frac{\sum_{i=1}^{n}(x_i - \bar{x})(y_i - \bar{y})}{\sum_{i=1}^{n}(x_i - \bar{x})^2}.$$

Similarly,

$$\frac{\partial LL(a_0, a_1)}{\partial a_0} = -\frac{1}{s^2}\sum_{i=1}^{n}(y_i - (a_0 + a_1 x_i)) = 0$$

when

$$a_0 = \frac{1}{n}\sum_{i=1}^{n} y_i - \frac{a_1}{n}\sum_{i=1}^{n} x_i = \bar{y} - a_1 \bar{x}.$$

In matrix form,

$$\begin{pmatrix} a_0 \\ a_1 \end{pmatrix} = (X^T X)^{-1} X^T Y,$$

where

$$X = \begin{pmatrix} 1 & x_1 \\ 1 & x_2 \\ \vdots & \vdots \\ 1 & x_n \end{pmatrix} \quad \text{and } Y = \begin{pmatrix} y_1 \\ y_2 \\ \vdots \\ y_n \end{pmatrix}.$$

Example 5.13. *Figure 5.4 shows five randomly generated interpolation points*

$$\begin{pmatrix} x & y \\ 3.33144 & 15.4277 \\ 0.42681 & -0.194758 \\ 1.10072 & 4.84892 \\ 1.91978 & 7.18721 \\ 2.29908 & 8.29164 \end{pmatrix}$$

with $(\bar{x}, \bar{y}) = (1.81557, 7.11213)$. *We use maximum likelihood to determine that*

$$a_1 = \frac{\sum_{i=1}^{n}(x_i - \bar{x})(y_i - \bar{y})}{\sum_{i=1}^{n}(x_i - \bar{x})^2} = 5.0076$$

$$a_0 = \bar{y} - a_1 \bar{x} = -1.9795$$

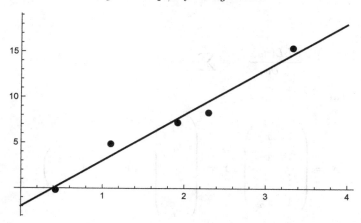

FIGURE 5.4
Data from Example 5.13 together with the maximum likelihood function,
$L(a_0, a_1) = -1.9795 + 5.0076x$.

A **Poisson regression model**, or log-linear model assumes that the random variables, Y_i, are independent identically distributed Poisson random variables, $Y_i \sim Poiss(\lambda)$.

Since $\lambda > 0$, we can determine $log(\lambda)$. We will assume that

$$log(\lambda) = a_0 + a_1 x.$$

$$p(y_i|x_i, a_0, a_1) = \prod_{i=1}^{n} e^{-\lambda} \frac{\lambda^{y_i}}{y_i!}$$

$$LL(a_0, a_1) = log\left(\prod_{i=1}^{n} e^{-\lambda} \frac{\lambda^{y_i}}{y_i!} \right)$$

$$= \sum_{i=1}^{n} -\lambda + y_i \cdot log(\lambda) - log(y_i!)$$

$$= \sum_{i=1}^{n} -e^{(a_0 + a_1 x_i)} + y_i \cdot (a_0 + a_1 x_i) - log(y_i!).$$

Taking partial derivatives,

$$\frac{\partial LL(a_0, a_1)}{\partial a_0} = \sum_{i=1}^{n} -e^{(a_0 + a_1 x_i)} + y_i = 0.$$

Similarly,

$$\frac{\partial LL(a_0,a_1)}{\partial a_1} = \sum_{i=1}^{n} -x_i e^{(a_0+a_1 x_i)} + x_i y_i.$$

In matrix form,

$$X = \begin{pmatrix} 1 & x_1 \\ 1 & x_2 \\ \vdots & \vdots \\ 1 & x_n \end{pmatrix}, \ Y = \begin{pmatrix} y_1 \\ y_2 \\ \vdots \\ y_n \end{pmatrix}, \text{ and } A = \begin{pmatrix} a_0 \\ a_1 \end{pmatrix},$$

the gradient

$$\nabla LL(A) = X^T Y - X^T e^{XA}.$$

There is not an algebraic solution to $\dfrac{\partial LL(a_0,a_1)}{\partial a_0} = 0$ and $\dfrac{\partial LL(a_0,a_1)}{\partial a_1} = 0.$

The numeric technique called the **Newton Raphson method** can be used to iteratively estimate the best choice for a_0 and a_1.

The Hessian matrix,

$$H_{LL}(A) = \begin{pmatrix} \dfrac{\partial^2 LL}{\partial a_0^2} & \dfrac{\partial^2 LL}{\partial a_0 \partial a_1} \\[2mm] \dfrac{\partial^2 LL}{\partial a_1 \partial a_0} & \dfrac{\partial^2 LL}{\partial a_1^2} \end{pmatrix}$$

$$= \begin{pmatrix} \sum_{i=1}^{n} e^{(a_0+a_1 x_i)} & \sum_{i=1}^{n} x_i e^{(a_0+a_1 x_i)} \\ \sum_{i=1}^{n} x_i e^{(a_0+a_1 x_i)} & \sum_{i=1}^{n} x_i^2 e^{(a_0+a_1 x_i)} \end{pmatrix}$$

$$= X^T W X,$$

where W is the diagonal matrix

$$W = \begin{pmatrix} e^{(a_0+a_1 x_1)} & 0 & 0 & \cdots & 0 \\ 0 & e^{(a_0+a_1 x_2)} & 0 & \cdots & 0 \\ \vdots & \ddots & \ddots & & \vdots \\ 0 & 0 & 0 & \cdots & e^{(a_0+a_1 x_n)} \end{pmatrix}.$$

Using the Hessian matrix, the ith iterative estimate of A, $A^{(i)}$, is

$$A^{(i)} = A^{(i-1)} - H_{LL}(A^{(i-1)})^{-1}\nabla LL(A^{(i-1)}).$$

Example 5.14. *Poisson regression is helpful in modeling counts of data. Walt Disney World uses Big Data and Data Analytics to collect a vast amount of data on its customers. In 2013, Disney released the Magic Band bracelet, which communicates with sensors spread throughout the parks, generating huge amounts of data about the movements of each individual customer.*

On five random days and at a random time during the day, the park collected data on the number of guests that had arrived to ride the new roller coaster. Table 5.2 shows that data.

In order to help with queue planning, the park wishes to predict how many customers arrive at the ride per hour.

TABLE 5.2
Example of data related to number of customers arriving at a ride per hour.

X(in hrs)	Time	1	4	8.5	5	10
Y	Visitors	110	500	1240	800	1570

$$\bar{x} = 324, \bar{y} = 1344$$

We begin with an initial guess for A, $A^{(0)} = \begin{pmatrix} 5 \\ .25 \end{pmatrix}$.

Additionally,

$$\nabla LL(A) = \begin{pmatrix} 57.3015 \\ -687.279 \end{pmatrix}$$

$$H_{LL}(A) = \begin{pmatrix} 4162.7 & 33037.3 \\ 33037.3 & 290181 \end{pmatrix}.$$

Then,

$$A^{(1)} = A^{(0)} - H_{LL}(A^{(0)})^{-1}\nabla LL(A^{(0)})$$
$$= \{4.6623, 0.290816\}.$$

If this method is applied again $A^{(2)} = \{3.89028, 0.38231\}$. Applying what we learned to determine a Poisson regression model,

$$log(\lambda/T) = a_0 + a_1(x)$$
$$log(\lambda) - log(T) = a_0 + a_1(x)$$

Recall that the park wishes to predict how many customers arrive at the ride in a 1 hour period. Then

$$log(\lambda) - log(1) = 3.89028 + .013821x.$$

Thus, on average, the number of customers that arrive at the ride per hour is approximately

$$\lambda = 71.7071.$$

5.4 Gradient Descent Method

In Section 4.6, an interpolation technique, Least Squares Regression, is introduced with a goal to determine an interpolating polynomial satisfying

$$\hat{f}_n(x) = \sum_{i=0}^{n} a_i x^n, \ \hat{f}(x_j) = f(x_j), \ 0 \le j \le n,$$

that minimizes the sum of the squared error,

$$SSE = \sum_{j=0}^{n} (\hat{f}(x_j) - f(x_j))^2.$$

In this section, an iterative method is presented that attempts to minimize the sum of squared errors, SSE.

In previous sections of this chapter, we explore techniques that require that interpolants match the given data points. The technique presented in this section has more flexibility with the degree of the interpolating polynomials but may not match the data points.

Matching the data points can be an important quality; however, there are other behaviors of the data that are equally important to consider, such as minimizing a given **loss function**, which in this case is the SSE.

In Figure 5.6, a loss function is pictured. The **gradient** (also called the rate of change or first derivative) of the function is the rate at which the function is changing. Notice that at point a, the function is decreasing, thus the gradient is negative.

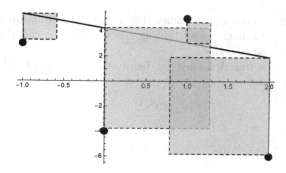

FIGURE 5.5
A visualization of the sum of squared errors, SSE.

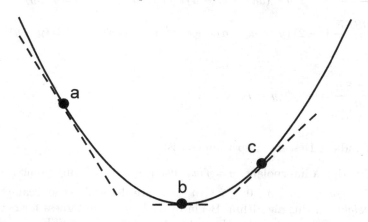

FIGURE 5.6
Visualization of the gradient of a loss function.

If the function is decreasing then the gradient is negative.

Similarly, notice that at point c in Figure 5.6, the function is changing in a positive direction. Most importantly, the rate at which the function is changing at a minimum (or maximum), the slope of the dashed line (on Figure 5.6), is zero.

If the function is increasing then the gradient is positive.

A function reaches an optimum value (either a minimum or maximum) when the gradient is zero.

Thus, one can determine the values that minimize a function by calculating the **gradient** of the function and finding the values, also called **parameters**, that optimize the loss function.

If **SSE** is the loss function then we denote the gradient of SSE with respect to the coefficient a_i as

$$\frac{\partial SSE}{\partial a_i}.$$

If $\hat{y} = f(\hat{x}) = a_0 + a_1 x + a_2 x^2 + \ldots a_n x^n$, then

$$SSE = (y - (a_0 + a_1 x + a_2 x^2 + \ldots a_n x^n))^2$$
$$\frac{\partial SSE}{\partial a_0} = -2(y - (a_0 + a_1 x + a_2 x^2 + \ldots a_n x^n)) = -2(y - \hat{y})$$
$$\frac{\partial SSE}{\partial a_1} = -2x(y - (a_0 + a_1 x + a_2 x^2 + \ldots a_n x^n)) = -2x(y - \hat{y})$$
$$\vdots$$
$$\frac{\partial SSE}{\partial a_2} = -2x^n(y - \hat{y}).$$

The Gradient Descent Algorithm consists of

1. Creating a interpolant, $\hat{y} = \hat{f}(x)$, using n interpolating points.

   The coefficients, a_i, $0 \leq i \leq n$, that are determined are called the *weights* in this algorithm. Begin by making a good guess for each of the coefficients and choosing a loss function, such as SSE.

2. For each coefficient, a_i, determine the gradient

$$\frac{\partial SSE}{\partial a_i} = \sum_{j=0}^{n} (y_j - \hat{y}_j)(x_j)^i.$$

3. Adjust the weights taking into consideration a learning rate, r,

$$\text{new } a_i = a_i + r\frac{\partial SSE}{\partial a_i}.$$

4. Repeat Steps 2-4 to update the coefficients and approach a minimum SSE.

Example 5.15. *In order to use the Gradient Descent method to fit a linear function $\hat{f}(x) = a_0 + a_1 x$ to the interpolation points*

$$\{(-1, 3), (0, -4), (1, 5), (2, -6)\},$$

we begin by choosing a random guess for the coefficients

$$a_0 = 5, a_1 = -.5.$$

The interpolant

$$f(\hat{x}) = 5 - .5x \text{ has a } SSE = 187.5.$$

y_i	$\hat{y}_i = a_0 + a_1 x_i$	$\dfrac{\partial SSE}{\partial a_0}$	$\dfrac{\partial SSE}{\partial a_1}$
3	5.5	5	-5
-4	5	18	0
5	4.5	-1	-1
-6	4	20	40
Total		42	36

If the learning rate is $r = .01$ then the new coefficients are

$$a_0 = 5 - .01 \cdot 42 = 4.58 \text{ and } a_1 = -.5 - .01 \cdot 36 = -.86.$$

The new interpolant $\hat{f}(x) = 4.58 - .86x$ has a $SSE = 159.708$.
Continuing in this fashion,

y_i	$\hat{y}_i = a_0 + a_1 x_i$	$\dfrac{\partial SSE}{\partial a_0}$	$\dfrac{\partial SSE}{\partial a_1}$
3	5.44	4.8	-4.88
-4	4.58	17.16	0
5	3.72	-2.56	-2.56
-6	2.86	17.72	35.44
Total		37.2	28

    The new model becomes $\hat{f}(x) = 4.208 - 1.14x$, with an $SSE = 139.47$.
Figure 5.7 shows the interpolation data and the interpolating function through
this process. We can continue in this manner; however, you may notice that a
linear interpolant may not be the best choice for this data.
    The next example extends this work, fitting a cubic interpolant to the given
interpolation points.

Example 5.16. *The cubic interpolation function $\hat{f}(x) = a_0 + a_1 x + a_2 x^2 + a_3 x^3$
requires a random guess for*

$$a_0 = -4, \ a_1 = 7, \ a_2 = 7, \text{ and } a_3 = -4.$$

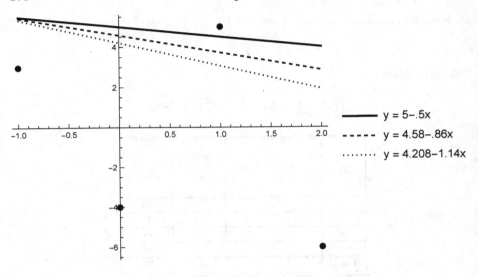

FIGURE 5.7
Example 5.15 data and linear interpolants for several iterations of gradient descent.

y_i	$\hat{f}(x_i)$	$\dfrac{\partial SSE}{\partial a_0}$	$\dfrac{\partial SSE}{\partial a_1}$	$\dfrac{\partial SSE}{\partial a_2}$	$\dfrac{\partial SSE}{\partial a_3}$
3	0	-6	6	-6	6
-4	-4	0	0	0	0
5	6	2	2	2	2
-6	6	24	48	96	192
Total		20	56	92	200

We will continue to use a learning rate $r = .01$; however, another small choice for r is also appropriate. With this choice for r, the updated interpolant becomes

$$\hat{f}(x) = -4.2 + 6.44x + 6.08x^2 - 6x^3$$

with $SSE = 154$. The interpolation functions along with the interpolants can be seen in Figure 5.8.

Gradient Descent methods can be written in terms of matrix system as well. If the interpolation function is an nth degree polynomial

$$\hat{y} = a_0 + a_1 x + \cdots + a_n x^n,$$

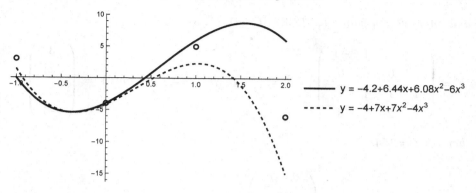

FIGURE 5.8
Example 5.15 data and cubic interpolants for several iterations of gradient descent.

then the SSE cost function can be written as

$$SSE = (XA - Y)^T(XA - Y), \text{ where}$$

$$X = \begin{pmatrix} 1 & x_0 & x_0^2 & \cdots & x_0^n \\ 1 & x_1 & x_1^2 & \cdots & x_1^n \\ \vdots & \ddots & \ddots & \ddots & \vdots \\ 1 & x_n & x_n^2 & \cdots & x_n^n \end{pmatrix}, \quad A = \begin{pmatrix} a_0 \\ a_1 \\ \vdots \\ a_n \end{pmatrix},$$

$$and \quad Y = \begin{pmatrix} y_0 \\ y_1 \\ \vdots \\ y_n \end{pmatrix}.$$

The gradients are

$$\frac{\partial SSE}{\partial A} = X^T(XA - Y),$$

and if r is the learning rate then

$$A_i^{(j+1)} = A_i^{(j)} - r\frac{\partial SSE}{\partial a_i},$$

where $A_i^{(j)}$ is the ith entry of the jth iteration of A.

Example 5.17. *Revisiting Example 5.16, with interpolant*

$$\hat{y} = a_0 + a_1 x + a_2 x^2 + a_3 x^3,$$

and interpolation points $\{(-1, 3), (0, -4), (1, 5), (2, -6)\}$,

$$X = \begin{pmatrix} 1 & -1 & 1 & -1 \\ 1 & 0 & 0 & 0 \\ 1 & 1 & 1 & 1 \\ 1 & 2 & 4 & 8 \end{pmatrix}, A^{(0)} = \begin{pmatrix} -4 & 7 & 7 & -4 \end{pmatrix}, \text{ and } Y = \begin{pmatrix} 3 \\ -4 \\ 5 \\ -6 \end{pmatrix}.$$

then $SSE = 154$.

$$\begin{pmatrix} \dfrac{\partial SSE}{\partial a_0} \\[2ex] \dfrac{\partial SSE}{\partial a_1} \\[2ex] \dfrac{\partial SSE}{\partial a_2} \\[2ex] \dfrac{\partial SSE}{\partial a_3} \end{pmatrix} = \begin{pmatrix} 20 \\ 56 \\ 92 \\ 200 \end{pmatrix}.$$

Thus if the learning rate $r = .01$,

$$A^{(1)} = \begin{pmatrix} -4 & 7 & 7 & -4 \end{pmatrix} - .01 \begin{pmatrix} 20 \\ 56 \\ 92 \\ 200 \end{pmatrix} = \begin{pmatrix} -4.2 \\ 6.44 \\ 6.08 \\ -6. \end{pmatrix}.$$

As you can tell from Examples 5.15 and 5.16, the Gradient Descent method can be computational expensive. That is, that for each iteration of the Gradient Descent method, computations have to be made on each data point.

In traditional gradient descent methods, the gradient (or derivative) of the loss function must be calculated for each interpolating point. This can be computationally hard. For example, in Example 5.16, there are only six data points with four partial derivatives in one iteration. After just four interactions, there are almost 100 calculations. If a dataset contained 10,000 points and one wished to use gradient descent to model the data with a cubic polynomials, each iteration would take 40,000 calculations.

Stochastic Gradient Descent is one way to make computational improvements to these methods. When employing Stochastic Gradient Descent, the gradient is calculated, on each iterative step, at a random point or small random set of points called a **mini-batch**.

Additionally, keep in mind that depending on the loss function, the gradient descent method is also not guaranteed to converge, so it is important to think carefully about both the learning rate, r, and the loss function that you are using. Figure 5.9 shows both an example in which the method converges and diverges.

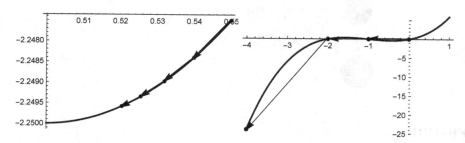

FIGURE 5.9
Examples of the gradient descent method converging (left) and diverging (right).

The gradient descent method is just one method of neural network. In the next section, we explore the concepts of neural networks more broadly.

5.5 Introduction to Neural Networks

Neural networks are powerful tools in data science that focus on relationships in data. It is interesting to note that neural networks are based on the systems in the brain. In the brain, neurons receive information, or data, from other neurons. This system of neurons is connected by axons, where some neurons are more connected that other neurons to the network.

There are two general types of neural networks, **unsupervised neural networks** that try to predict an output given a set of input data and **supervised neural networks** where both the input data and output data is know. In a supervised neural network, the known output data can be used to determine the accuracy of the neural network.

In an unsupervised neural network, a neuron can be viewed as receiving a set of inputs, $\{x_1, x_2, \ldots, x_n\}$, to predict one output, y. Each input is associated with a weighted connection between it and the output. We will represent the matrix of weights as W. Data also naturally has bias or noise; we represent this bias with the vector b.

With just inputs and one output, the neural network, in its simplest form as it is visualized in Figure 5.10, is called a **perceptron**. The perceptron is not a very interesting case of a neural network as we can accomplish the same thing with the techniques provided in previous sections of this chapter, writing $y = WX$, where X incorporates the inputs and the constant, bias, term.

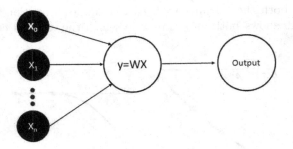

FIGURE 5.10
Visualization of a perceptron.

If $W = \{w_1, w_2, \ldots, w_n\}$ are the weights, then we can think of this as a linear combination of basis vectors, functions. For example, if $h_1(x) = 1$ and $h_2(x) = x$,
$$\hat{y} = w_1 h_1(x) + w_2 h_2(x)$$
is a linear model. Similarly, if one wished to use a quadratic function as the model, then the basis vectors would be
$$\{1, x, x^2\}.$$

The concept of neural networks becomes much more interesting when there is a **hidden layer**, S, adding to the complexity between the input and output values.

Like the gradient descent method, we start with a random guess for W and adjust the weights iteratively.

In addition, an **activation function**, can be incorporated such that
$$S = \phi(WX), \text{ and } y = \overline{W}S,$$
$\overline{W}$ is an updated weight matrix found after S is determined. There are a variety of activation techniques, which will be discussed in detail in Sections 5.5.2 and 5.5.3.

We begin with a simple example based on a binary decision, and then we will see how this applies to the interpolation problem.

Example 5.18. *Virginia is trying to decide whether to have a picnic with friends this weekend. She have two factors that you are considering when making this decision*

$$x_1 \rightarrow \text{ Will more than five friends be able to attend?}$$
$$x_2 \rightarrow \text{ Will the weather be pleasant?}$$

If we assign a false value 0 and true value 1, there are 4 possibilities of the input given these two factors

$$(0,0),(0,1),(1,0),(1,1)$$

In the first case $(0,0)$*, less than five of Virginia's friends can attend and the weather will not be pleasant.*

You may wish to put a weight to your two factors. For example, it may be more important that it does not rain then the number of friends that can attend.

Let's say that Virginia assigns weights $w_1 = 1$ *and* $w_2 = 4$ *to factors* x_1 *and* x_2 *respectively and a threshold value for which she will decide to hold the picnic equal to 6. There may also be a constant value, b, associated with the decision.*

$$z = w_1 x_1 + w_2 x_2 + b = x_2 + 4x_2 + 3.$$

Virginia then plugs in each of her input vectors

$$(0,0) \rightarrow 3$$
$$(0,1) \rightarrow 7 \rightarrow PICNIC$$
$$(1,0) \rightarrow 4$$
$$(1,1) \rightarrow 8 \rightarrow PICNIC$$

Machine learning works to adjust the weights and threshold in order to minimize errors. In addition, when using neural networks for an interpolation problem, typically a smoothing activation function, $\phi(WX)$, is applied at each step.

5.5.1 The Learning Process

The most important aspect of neural networks is the ability to learn. In the context of interpolation, this can be interpreted such that at each step the weights, w_i, $i = 1, \ldots n$, and $\bar{w}_j, j = 1, \ldots m$, are adjusted to minimize the error.

After applying the activation function, $\phi(S) = \phi(WX)$, the Gradient Descent method can be applied to find optimal weights that minimize the SSE or any other cost function.

When an activation function is applied with learning rate, r, and the SSE as the cost function,

$$\text{new } w_i = w_i - r \frac{\partial SSE}{\partial w_i}.$$

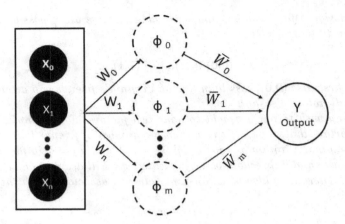

FIGURE 5.11
Example of General Neural Network with Activation Function.

5.5.2 Sigmoid Activation Functions

There are many different activation functions that you can use for $\phi(x)$. The activation function, or transfer function, is the transformation function that creates the output nodes. Thus, the choice of activation function can have a large impact on the performance of the neural network.

Activation functions are typically differentiable. This is because, in a learning process integrating a technique such as gradient descent, occurs after the activation function is applied. We will introduce four possible activation functions here and show you how they perform.

Typical smoothing functions from **regularization** can be used for $\phi(x)$. The first activation function of interest is the **SoftMax function**, where

$$\phi(x_k) = \frac{e^{x_k}}{\sum_{i=1}^{n} e^{x_i}},$$

and the partial derivative of $\phi(x_k)$ with respect to x_k is

$$\frac{\partial \phi(x_k)}{\partial x_k} = \frac{e^{x_k} \sum_{i=1, i \neq k}^{n} e^{x_i}}{\left(\sum_{i=1}^{n} e^{x_i}\right)^2}.$$

Example 5.19. *Let's imagine that we have collected data on the effectiveness of a 2 dose COVID vaccine. Those who did not have either dose had no immunity to the disease, those who got 1 dose of the vaccine were 30% immune to the newest strain of COVID, and those who received 2 doses of the vaccine were 90% immune.*

We would like to use this information to determine if a future booster would be effective.

Here we have three input values. For this example, we will incorporate the SoftMax activation function with two different initial weight functions, whose coefficients are located in the columns of W, prior to the activation (any number of weight functions, and corresponding activation functions, could be incorporated).

Let $W = \begin{pmatrix} 0.25 & 0.1 \\ 0.25 & 0.9 \end{pmatrix}$,

$$X = \begin{pmatrix} 1 & 0 \\ 1 & 1 \\ 1 & 2 \end{pmatrix}, Y = \begin{pmatrix} 0 \\ .3 \\ .9 \end{pmatrix}.$$

Then

$$(\phi(XW))^T = \left(\frac{e^{XW}}{\sum_{j=1}^{3} e^{(XW)_{ij}}} \right)^T$$

$$= \begin{pmatrix} 0.254275 & 0.105161 \\ 0.326496 & 0.258654 \\ 0.419229 & 0.636186 \end{pmatrix}.$$

We also need initial weights, $\overline{W}$, to put the activation functions together. For simplicity we will choose $\overline{W} = \{.5, .5\}$,

$$(\phi(XW))^T(\overline{W}) = \begin{pmatrix} 0.179718 \\ 0.292575 \\ 0.527707 \end{pmatrix}.$$

Back propagation, using a technique such as the gradient descent method, is then used to optimize the values in $\overline{W}$ and \or W. If one wishes to continue to use the initial guess for values in W then use

$$\frac{\partial SSE}{\partial \overline{W}} = -2(Y - \phi(XW)\overline{W})\phi(XW)$$

iteratively to minimize SSE while updating values of $\overline{W}$. Many iterations arrive at a steady state of approximately

$$\overline{W} = \{-0.448, 1.69\},$$

$$\phi(XW)\overline{W} = \begin{pmatrix} 0.252475 & 0.105161 \\ 0.326496 & 0.258654 \\ 0.419229 & 0.636186 \end{pmatrix} \begin{pmatrix} -0.448 \\ 1.69 \end{pmatrix}$$

predicting that the vaccine is 6.3%, 29%, *and* 88% *effective for 0, 1, and 2 doses respectively.*

The model itself is

$$\phi(XW)\overline{W} = \frac{-.448e^{.25+.25x}}{\sum_{i=1}^{3} e^{.25+.25x_i}} + \frac{1.69e^{.1+.9x}}{\sum_{i=1}^{3} e^{.1+.9x_i}}$$

If in addition, one wishes to iteratively change W *with gradient descent then note that*

$$\frac{\partial SSE}{\partial W} = -2(Y - \phi(XW)\overline{W})X\overline{W}\phi'(XW).$$

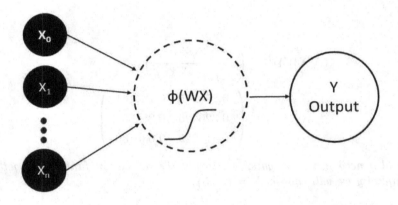

FIGURE 5.12
Neural network with sigmoid activation function.

Another common activation function is called the **logistic sigmoid neuron function**.

$$\phi(X) = \frac{1}{1 + e^{-(WX)}}.$$

The benefit of using the logistic sigmoid neuron function is that all output values will be between 0 and 1, and thus are good predictors related to probability.

Figure 5.12 gives you a general idea of how the sigmoid activation function acts within a neural network, and Figure 5.13 shows an example of a logistic sigmoid neuron function.

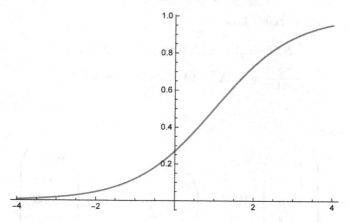

FIGURE 5.13
Logistic function centered at $x = 1$.

Example 5.20. *For the data set*

$$
\begin{pmatrix}
x & y \\
1 & 0.795 \\
2 & 0.814 \\
3 & 0.868 \\
4 & 0.66 \\
5 & 0.29
\end{pmatrix},
$$

we could start with initial weights $a_0 = 1$ and $a_1 = -0.1$ with

$$S = WX = 1 - 0.1x,$$

Note that if $\phi(x) = \frac{1}{1+e^{-x}}$ then

$$\phi'(x) = \frac{e^{-x}}{(1 + e^{-x})^2}.$$

In the learning process, when $\hat{y} = a_0 + a_1 x + a_2 x^2 + \cdots + a_n x^n$,

$$
X = \begin{pmatrix}
1 & x_0 & x_0^2 & \cdots & x_0^n \\
1 & x_1 & x_1^2 & \cdots & x_1^n \\
\vdots & \ddots & \ddots & \ddots & \ddots \\
1 & x_n & x_m^2 & \cdots & x_m^n
\end{pmatrix},
W = \begin{pmatrix}
a_0 \\
a_1 \\
\vdots \\
a_n
\end{pmatrix},
and\ Y = \begin{pmatrix}
y_0 \\
y_1 \\
\vdots \\
y_n
\end{pmatrix},
$$

and $\phi(x)$ is a differentiable function,

$$SSE = (Y - \phi(XW))^T(Y - \phi(XW))$$
$$\frac{\partial SSE}{\partial W} = -2X^T\phi'(XW) \circ (Y - \phi(XW)),$$

In this example,

$$X = \begin{pmatrix} 1 & 1 \\ 1 & 2 \\ 1 & 3 \\ 1 & 4 \\ 1 & 5 \end{pmatrix}, \; W^{(0)} = \begin{pmatrix} 1 \\ -.1 \end{pmatrix}, \; and \; Y = \begin{pmatrix} 0.795 \\ 0.814 \\ 0.868 \\ 0.66 \\ 0.29 \end{pmatrix},$$

$$SSE = 0.173107$$
$$\frac{\partial SSE}{\partial a_0} = -0.0265121$$
$$\frac{\partial SSE}{\partial a_1} = 0.348567.$$

In this process, a learning rate, r, must be chosen. If $r = .05$ then

$$W^{(1)} = W^{(0)} - r\frac{\partial SSE}{\partial W} = \begin{pmatrix} 1 \\ -.1 \end{pmatrix} - .05 \begin{pmatrix} -0.0265121 \\ 0.348567 \end{pmatrix} = \begin{pmatrix} 1.00133 \\ -0.117428 \end{pmatrix}.$$

Figure 5.14 shows S, $\phi(S)$, and the original data. Notice that in Example 5.5.2, weights were imposed on both the original inputs, X, and the activation functions, whereas in Example 5.5.2, weights were imposed on only the inputs. Weights on inputs and/or activation functions can be used to best model scenarios.

Another commonly used activation function is

$$\phi(x) = \tanh(x),$$

the **hyperbolic tangent** activation function.

In this case,

$$\phi'(x) = sech^2(x).$$

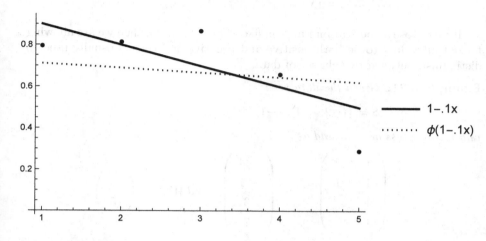

FIGURE 5.14
$S, \phi(S)$.

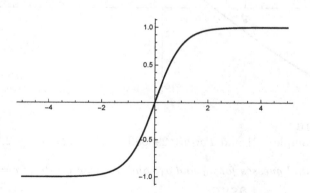

FIGURE 5.15
$\phi(x) = \tanh(x) = \frac{2}{1+e^{-2x}} - 1.$

If $\hat{Y} = \phi(XW)$,

$$\frac{\partial SSE}{\partial W} = -2X^T sech^2(XW) \circ (Y - \tanh(XW))$$

If we choose an activation function like $\phi(x) = tanh(x)$, then we are allowing for output values to be both positive and negative and thus is usually used to distinguish between two classes of data.

Example 5.21. *Given the data set*

$$S = \{(-2, -.4), (-1, -.1), (1, .3), (2, .7)\},$$

and initial guess $a_0 = .3$ and $a_1 = .5$,

$$X = \begin{pmatrix} 1 & -2 \\ 1 & -1 \\ 1 & 1 \\ 1 & 2 \end{pmatrix}, Y = \begin{pmatrix} -.4 \\ -.1 \\ .3 \\ .7 \end{pmatrix}, \text{ and } W^{(0)} = \begin{pmatrix} .3 \\ .5 \end{pmatrix}.$$

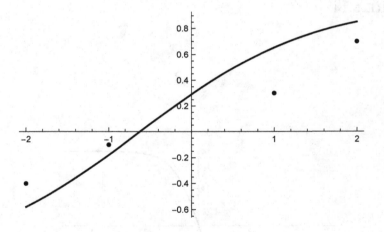

FIGURE 5.16
Data from Example 5.21 and $Tanh(0.298689 + 0.461612x)$ on $[-2, 2]$.

With these initial guesses for a_0 and a_1, and a learning rate of $r = .03$,

$$\frac{\partial SSE}{\partial W} = \{0.0436947, 1.27961\}$$

$$W^{(1)} = \begin{pmatrix} 0.298689 \\ 0.461612 \end{pmatrix}$$

The result $\hat{Y} = \phi(XW)$ can be seen in Figure 5.16.

Finally, one of the most widely used activation functions is the **rectified linear unit** function, **ReLU**. Defined as

$$\phi(x) = \begin{cases} 0, & \text{if } x < 0, \\ x, & \text{otherwise.} \end{cases}$$

Notice that the ReLu activation function is not differentiable at $x = 0$ and

$$\phi'(x) = \begin{cases} 0, & \text{if } x < 0, \\ \text{undefined}, & \text{if } x = 0, \\ 1, & \text{if } x > 0. \end{cases}$$

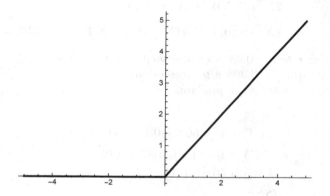

FIGURE 5.17
ReLu activation function.

The ReLu activation function is the most commonly used activation function because of it's accuracy while remaining easy to use and train.

Additionally, the ReLu activation function is used to manage the interaction between input, dependent, variables.

For example, if there are two input variables, x_1 and x_2, an activation function ϕ acts on the input variables, $\phi(w_1x_1 + w_2x_2)$. Let's assume that $w_1 = w_2 = 1$, then

$$\phi(w_1x_1 + w_2x_2) = \begin{cases} 0, & x_1 < |x_2| \\ x_1 + x_2, & x_1 \geq |x_2| \end{cases}.$$

Note that if both $x_1 > |x_2| > 0$ then both input variables contribute to the output.

For example, if $x_1 = 10$ and $x_2 = -2$ then $\phi(w_1x_1 + w_2x_2) = 8$. However, if either variable is a significantly larger negative number, then neither variable unduly influences the output. If $x_1 = 10$ and $x_2 = -1000$ then $\phi(w_1x_1+w_2x_2) = 0$.

Example 5.22. *In Example 5.21, we implemented the hyperbolic tangent activation function with initial guess $a_0 = .3$ and $a_1 = .5$ on*

$$\{(-2, -.4), (-1, -.1), (1, .3), (2, .7)\},$$

Let's now see what happens when we use the ReLu activation function, $\phi(x)$, on this same data set.

$$XW = \{-0.7, -0.2, 0.8, 1.3\},$$
$$\phi(XW) = \hat{Y} = \{0, 0, 0.8, 1.3\},$$
$$\phi'(XW) = \{0, 0, 1, 1\},$$
$$\frac{\partial SSE}{\partial W} = -2X^T(\phi'(XW) \circ (Y - \phi(XW)))$$
$$= -2X^T(\{0, 0, 1, 1\} \circ (Y - \{0, 0, 0.8, 1.3\})) = \{2.2, 3.4\}.$$

If a learning rate $r = 0.03$ is chosen then $a_0 = 0.3 - 0.03 \cdot (2.2) = 0.234$ and $a_1 = 0.5 - 0.03 \cdot (3.4) = 0.398$ after one iteration.
An additional iteration would produce,

$$XW = \{-0.562, -0.164, 0.632, 1.03\},$$
$$\phi(XW) = \hat{Y} = \{0, 0, 0.632, 1.03\},$$
$$\phi'(XW) = \{0, 0, 1, 1\},$$
$$\frac{\partial SSE}{\partial W} = \{1.324, 1.984\},$$

With the same learning rate, $a_0 = 0.19428$ and $a_1 = 0.33848$. Figure 5.18 shows a visualization of each iteration and the original data.

Next, let's see how to apply these ideas to a regression problem involving more than one input variable.

TABLE 5.3
Energy data [40, 41].

Country	Normalized kwh of Energy consumed per person(2019)	Normalized % of Energy from Fossil Fuel(2019)	Normalized GDP per capita in USD(2019)
USA	1.08354	0.525247	1.06688
Brazil	-0.826228	0.587071	-0.97821
China	0.616389	-1.49447	0.62683
Sweden	-0.873701	0.382148	-0.715505

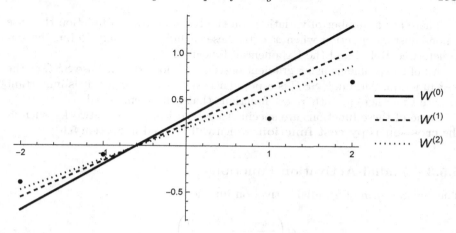

FIGURE 5.18
Iterations of ReLu activation in Example 5.22.

Example 5.23. *We will let A represents the normalized amount of energy consumed per person and B is the normalized % of energy from fossil fuel. Assume that $w_0 = .3$, $w_1 = .7$, and $w_2 = .1$ and we wish to predict, Y, the normalized per capita GDP, where*

$$X = \begin{pmatrix} 1 & 1.08354 & 0.525247 \\ 1 & -0.826228 & 0.587071 \\ 1 & 0.616389 & -1.49447 \\ 1 & -0.873701 & 0.382148 \end{pmatrix}, Y = \begin{pmatrix} 1.06688 \\ -0.97821 \\ 0.62683 \\ -0.715505 \end{pmatrix},$$

$$\hat{Y} = XW = \hat{Y} = w_0 + w_1 A + w_2 B = \{1.111, -0.219652, 0.582025, -0.273376\}$$

and $SSE = 0.774842$.

We display one iteration using ReLu, $\phi(XW)$, with SSE as the cost function.

$$\phi(XW) = \phi(\{1.111, -0.219652, 0.582025, -0.273376\})$$
$$= \{1.111, 0, 0.582025, 0\}$$
$$\phi'(XW) = \{1, 0, 1, 0\}$$
$$\frac{\partial SSE}{\partial W} = -2X^T(\phi(XW) \circ (Y - \phi(XW)))$$
$$= \{-0.001364, 0.0403832, 0.180269\}$$

With a learning rate of $r = 0.03$, new values of w_0, w_1, and w_2 respectively are 0.299731, 0.696355, and 0.0853578.

There are a number of variations on the ReLu activation function that use a non-constant function when $x < 0$, these include the Leaky ReLu, the Parameterised ReLu, and the Exponential Linear Unit.

All of the problems using gradient descent methods thus far use SSE as the cost function. Although SSE is a popular cost function to use, it is important to note that there are other cost functions that can be employed.

Some of these functions are specific to categorical neural networks, such as the **cross-entropy cost function**, which will be used in Section 5.6.

5.5.3   Radial Activation Functions

The general form of a radial activation function is

$$\phi\left(\frac{(x-c)^T(x-c)}{r^2}\right),$$

where c is a centering parameter and r is a scalar radius. Notice that when using radial activation functions, the function $\phi(x)$ acts on $||x-c||$.

Figure 5.20 shows some examples of radial activation functions.

One standard smoothing function, or **convolution** function, used in regularization is a Gaussian activation function, or **Gaussian kernel**,

$$\phi(x) = e^{\left(\frac{-(x-c_j)^T(x-c_j)}{\sigma^2}\right)}.$$

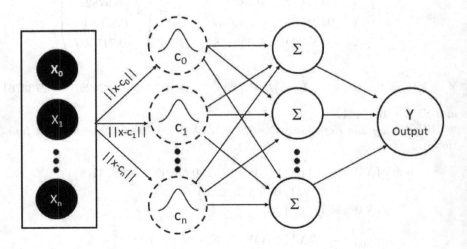

FIGURE 5.19
Visualization of implementation of radial activation functions.

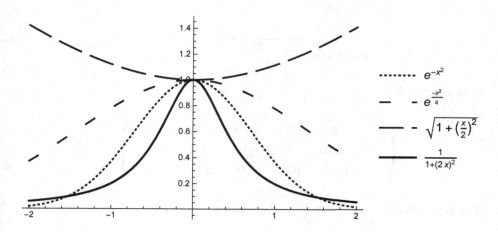

FIGURE 5.20
Examples of Radial Activation Functions.

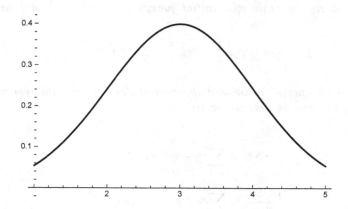

FIGURE 5.21
Example of Gaussian kernel with radius 2 centered at 3.

Example 5.24. *Given two input variables $X_1 \sim N(0,1)$ and $X_2 \sim N(0, .25)$, we wish to determine a Gaussian activation function,*

$$\phi_j(x) = e^{-\dfrac{(x - c_j)^T (x - c_j)}{2\sigma_j^2}}, j = 1, 2,$$

to model the data

$$\begin{pmatrix} x_1 & -2.1 & -0.92 & 0.08 & 0.96 & 2.1 \\ x_2 & -0.94 & -1 & -1.1 & -1.08 & -0.96 \\ y & 0.08 & 0.32 & 0.46 & 0.18 & -0.02 \end{pmatrix}.$$

The goal is to determine the optimal w_1 and w_2 such that

$$\phi(x) = \sum_{j=1}^{2} w_j \phi_j(x),$$

minimizes the sum of squared errors, SSE.

In this example, we are given c_j and σ_j^2 for $j = 1, 2$; however, one might have to use some sort of clustering technique to determine the centers and spreads of the input variables.

For simplicity, we begin with initial guess $w_1^{(0)} = w_2^{(0)} = \frac{1}{2}$ with activation function

$$\phi^{(0)}(x) = \frac{1}{2}e^{-\frac{(x-0)^2}{2}} + \frac{1}{2}e^{-2(x-0)^2}.$$

The sum of squared error and the partial derivatives with respect to the weights, w_1 and w_2, at iteration k are

$$SSE = \sum_{i=1}^{4}(y - \phi^{(k)}(x))^2$$

$$\frac{\partial SSE}{\partial w_1} = \sum_{i=1}^{4} -2(y - \phi^{(k)}(x))e^{-\frac{x^2}{2}}$$

$$\frac{\partial SSE}{\partial w_2} = \sum_{i=1}^{4} -2(y - \phi^{(k)}(x))e^{-2x^2}$$

TABLE 5.4

Estimations of w_1 and w_2 in the iterative process with $r = .03$.

Iteration	w_1	w_2	SSE	$r \cdot \frac{\partial SSE}{\partial w_1}$	$r \cdot \frac{\partial SSE}{\partial w_2}$
0	.5	.5	0.074	0.0163	0.0042
1	0.4837	0.4958	0.0649	0.014	0.0039
2	0.4697	0.4919	0.0580	0.013	0.0037
3	0.4567	0.4882	0.0524	0.011	0.0034
4	0.4457	0.4848	0.0481	0.010	0.0032

Example 5.24 shows how to apply a Gaussian radial activation function when the centers and spreads of data are known. In most data situations, the centers are unknown. We will next revisit this data problem from the Case Study in 3.8.

Example 5.25. *Recall that in the Case Study in 3.8, we were presented with two small samples of facial pixel data, where each column of X_1 and X_2 represents an image,*

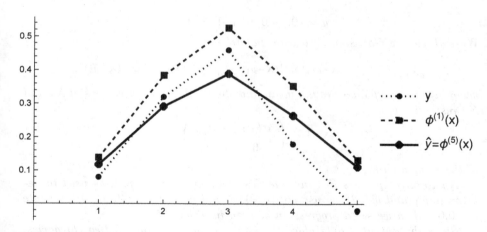

FIGURE 5.22

Iterations of a radial activation in Example 5.24.

$$X_1 = \begin{pmatrix} 97 & 125 & 106 & 172 \\ 96 & 93 & 100 & 185 \\ 81 & 72 & 93 & 200 \\ 87 & 70 & 103 & 171 \\ 95 & 54 & 97 & 187 \\ 86 & 55 & 88 & 203 \\ 48 & 58 & 100 & 169 \\ 67 & 47 & 96 & 187 \\ 90 & 51 & 87 & 203 \end{pmatrix} , X_2 = \begin{pmatrix} 158 & 216 & 191 & 91 \\ 153 & 218 & 218 & 94 \\ 147 & 218 & 244 & 98 \\ 156 & 215 & 193 & 86 \\ 154 & 212 & 225 & 92 \\ 151 & 209 & 227 & 98 \\ 159 & 207 & 178 & 82 \\ 158 & 201 & 203 & 89 \\ 156 & 195 & 215 & 97 \end{pmatrix} ,$$

and two additional unclear facial images represented by the pixel data in f_1 and f_2,

$$f_1 = \{238, 239, 240, 235, 236, 237, 232, 233, 234\}$$
$$f_2 = \{193, 193, 193, 193, 192, 192, 192, 192, 191\}.$$

The goal was to put f_1 and f_2 into one of the two categories, X_1 or X_2. We will begin by putting the eight images into two categories $X_1 \to 0$ and $X_2 \to 1$,

$$y = \{0, 0, 0, 0, 1, 1, 1, 1\}.$$

We will use the Gaussian activation function,

$$\phi(x) = w_1 e^{-\frac{1}{2}(X-\mu_1)^T \Sigma^{-1}(X-\mu_1)} + w_2 e^{-\frac{1}{2}(X-\mu_2)^T \Sigma^{-1}(X-\mu_2)},$$

where μ_1, σ_1 and μ_2, σ_2 are the mean and standard deviation vectors of X_1 and X_2 respectively,

$$\Sigma^{-1} = \begin{pmatrix} \sigma_1^T \sigma_1 & 0 \\ 0 & \sigma_2^T \sigma_2 \end{pmatrix},$$

of the sample data.

  An iterative process like the one described in Example 5.24 is used to determine w_1 and w_2. If we use $w_1 = -0.08$ and $w_2 = 0.9$ with a learning rate $r = 0.05$, then we see a progression as seen in Table 5.5.

  Since the output should place the blurry image into one of two categories, X_1 represented with a 0 or X_2 represented by a 1, it may be more practical to determine a final rounded output, where $\lfloor x \rceil$ represents rounding x to the nearest integer.

  Using $w_1 = -0.0820965$ and $w_2 = 0.895667$,

$$\phi(f_1) = \lfloor 0.425659 \rceil \to Category\ 0$$
$$\phi(f_2) = \lfloor 0.780331 \rceil \to Category\ 1$$

TABLE 5.5

Estimations of w_1 and w_2 in the iterative process with $r = .05$.

w_1	w_2	SSE	$\dfrac{\partial SSE}{\partial w_1}$	$\dfrac{\partial SSE}{\partial w_2}$
-0.08	0.9	1.60536	0.0373608	0.0565986
-0.081868	0.89717	1.6052	0.00857073	0.0275096
-0.0822966	0.895795	1.60517	-0.00176823	0.0158361
-0.0822082	0.895003	1.60516	-0.00516558	0.0108391
-0.0819499	0.894461	1.60515	-0.00598346	0.0084358

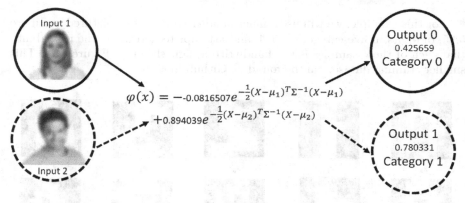

FIGURE 5.23

Visualization of using radial activation to categorize data.

In Chapter 6, we explore other iterative processes that are used for categorical decisions and regression analysis.

5.6   CASE STUDY: Handwriting Digit Recognition

The MNIST is a training set for handwritten digit recognition [44]. This training set consists of 60000 examples in 10 categories, 0–9. In this example, we show a small example but encourage the reader to explore the larger dataset by implementing the R code found at Github link 34 or the Python code at Github link 32.

Assume that you have a handwritten digit such as the one pictured in Figure 5.24 and you wish to determine what digit the handwriting represents.

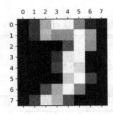

FIGURE 5.24
Sample handwritten digit to identify.

For this example, we will use a much smaller training set, pictured in Figure 5.25, of digits in five categories, 0–4, and attempt to determine the probability that each of these samples is the handwritten digit shown in Figure 5.24. This smaller training dataset can be found at Github link 36.

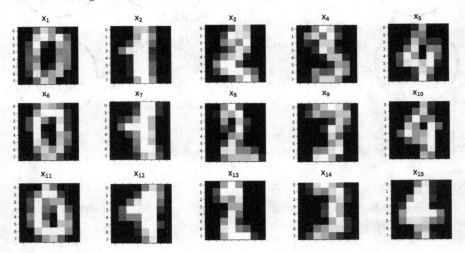

FIGURE 5.25
Small sample of digits from the training database.

Begin by assuming initial weights for each of the 15 inputs in the sample training set. We have chosen to make each sample of training data equally likely and thus equal weights, which is not a necessary assumption.

$$\{w_1, w_2, \ldots, w_{15}\} = \left\{\frac{1}{15}, \frac{1}{15}, \ldots, \frac{1}{15}\right\}$$

such that $\hat{y} = w_1 x_1 + w_2 x_2 + \cdots + w_{15} x_{15} = XW$.

We then apply an activation function, $\phi(x)$. In this case, we employ a neural network analysis, with the hyperbolic tangent activation function and traditional SSE cost function.

Choosing a learning rate, r, for example $r = .03$ to determine

$$\frac{\partial SSE}{\partial W} = -2X^T(\phi'(XW) \circ (Y - \phi(XW)))$$

$$W^{(i+1)} = W^{(i)} - r \cdot \frac{\partial SSE}{\partial W}.$$

Additionally, if the weights, w, are a percent vector, set

$$w_i = \frac{w_i}{\sum_{i=1}^{15} w_i}.$$

If

$$\phi(X) = tanh(x) \text{ and } \phi'(x) = sech^2(x),$$

then after twenty iterations of the learning algorithm,

$$w = \begin{pmatrix} 0.08427255 \\ -0.00643189 \\ -0.02651266 \\ \mathbf{0.12066096} \\ -0.00716325 \\ 0.03448799 \\ 0.00267105 \\ 0.08602818 \\ \mathbf{0.23687438} \\ -0.0082189 \\ -0.01507112 \\ 0.06937072 \\ 0.13791523 \\ \mathbf{0.29641327} \\ -0.0052965 \end{pmatrix},$$

and thus it is most probable that our digit is x_4, x_9, or x_{14}. Predicting that the given image is the digit 3.

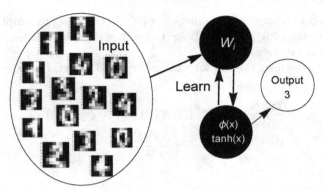

FIGURE 5.26
Diagram of the Neural Network Process.

Since we are dealing with choosing a category for our input image, we can also choose to employ the **categorical cross-entropy cost function**, instead of SSE.

$$C(X) = -\sum_{j=1}^{n} y_j \cdot log(\phi(x_j w_j)) = -Y log(\phi(XW))$$

with partial derivatives,

$$\frac{\partial C}{\partial W} = -\frac{X^T Y}{\phi(XW)} \phi'(XW).$$

If the activation function is the softmax function, $\phi(x_j) = \dfrac{e^j}{\sum_{k=1}^{n} e_k}$, then

$$\frac{\partial C}{\partial w_k} = -\sum_{j=1}^{n} \frac{y_j}{\phi(x_j w_j)} \frac{\partial \phi(x_j w_j)}{\partial w_k}$$

$$= \frac{-x_k y_k \phi(x_k w_k)(1 - \phi(x_k w_k))}{\phi(x_k w_k)} + \sum_{j \neq k}^{n} \frac{x_j y_j \phi(x_j w_j)\phi(x_k w_k)}{\phi(x_j w_j)}$$

$$= -x_k y_k (1 - \phi(x_k w_k)) + \sum_{j \neq k} x_j y_j \phi(x_k w_k)$$

$$= -x_k y_k + \sum_{j=1}^{n} x_j y_j \phi(x_k w_k).$$

We will choose y to be a vector where $y_k = 1$ and $y_j = 0$ for all $j \neq k, 1 \leq k \leq n$. Thus,

$$\frac{\partial C}{\partial w_k} = -x_k y_k + x_k \phi(x_k w_k) = x_k(-y_k + \phi(x_k w_k)).$$

Cross-entropy is computed for each value of k, noting that the smaller the entropy the more similar the input is to y_k.

In this example, with the initial values for W,

$$XW(0) = 0, \; XW(1) = \frac{1}{5}, \; XW(2) = \frac{11}{3}, \; \text{and} \; \sum_{j=1}^{n} XW(j) = \frac{4568}{15}.$$

Applying the softmax activation function,

$$\phi(XW)(0) = \frac{15e^0}{4568}, \; \phi(XW)(1) = \frac{15e^{\frac{1}{5}}}{4568}, \; \text{and} \; \phi(XW)(1) = \frac{15e^{\frac{11}{3}}}{4568}.$$

In this problem, we need to calculate the cross-entropy for all 15 images in the training set; however, if there were only these three images with $y_1 = \{1,0,0\}$, $y_2 = \{0,1,0\}$, and $y_2 = \{0,0,1\}$, the cross-entropy associated with y_1 would be

$$C = 1 \cdot log\left(\frac{15e^0}{4568}\right) + 0 \cdot log\left(\frac{15e^{\frac{1}{5}}}{4568}\right) + 0 \cdot log\left(\frac{15e^{\frac{11}{3}}}{4568}\right) \approx -5.71878.$$

Similarly, the cross-entropy associated with y_2 is

$$C = 0 \cdot log\left(\frac{15e^0}{4568}\right) + 1 \cdot log\left(\frac{15e^{\cdot 2}}{4568}\right) + 0 \cdot log\left(\frac{15e^{\frac{11}{3}}}{4568}\right) \approx -5.51878.$$

and that associated with y_3 is

$$C = 0 \cdot log\left(\frac{15e^0}{4568}\right) + 1 \cdot log\left(\frac{15e^{\cdot 2}}{4568}\right) + 0 \cdot log\left(\frac{15e^{\frac{11}{3}}}{4568}\right) \approx -2.05211.$$

then the image associated with y_3 is most similar to the input image x.

5.7 CASE STUDY: Poisson Regression and COVID Counts

Suppose that an experiment consists of counting the number of confirmed COVID-19 cases each day over a period of time, T.

Define Y_t as the random variable representing the total number of newly confirmed positive cases on day t, where $t = 1, 2, \ldots, T$. Thus the probability density function of Y_t is

$$L(Y_t|\ t, a_0, a_1) = \prod_{t=1}^{T} \frac{(a_0 + a_1 t)^{Y_t}}{Y_t!} e^{-(a_0 + a_1 t)},$$

where the mean

$$E(Y_t|\ t, a_0, a_1) = \lambda = \ln(a_0 + a_1 t).$$

The data for this Case Study is from the open source data site provided by the Ontario government on over 500000 confirmed cases of COVID-19 in Ontario in 2020 and 2021 [27]. For this case study, a small sample of this data is displayed. The reader is encouraged to explore this data set further.

TABLE 5.6
COVID-19 cases in Ontario, Canada in April 2020.

Date	4/13	4/14	4/15	4/16	4/17	4/18	4/19	4/20
Count	607	576	638	615	655	536	358	539
Date	4/21	4/22	4/23	4/24	4/25	4/26	4/27	4/28
Count	432	460	445	428	448	367	470	401

We will model a log-linear model to this data and start with an initial guess of the coefficients of $A^{(0)} = \{5.5, .4\}$.

With this guess,

$$LL(A^{(0)}) = \{-372768., -5.45396 \times 10^6\}$$

$$H(A^{(0)}) = \begin{pmatrix} 445954. & 6.2404 \times 10^6 \\ 6.2404 \times 10^6 & 8.98843 \times 10^7 \end{pmatrix}$$

$$A^{(1)} = \{5.03667, 0.49285\}$$

In this case,

$$\lambda = 5.03667 + 0.49285t.$$

This information can also be used to determine the $R0$ in mid April 2020 in Ontario, Canada.

5.8   Exercises

1. Given $y = \begin{pmatrix} y_1 \\ y_2 \end{pmatrix}$ $x = \begin{pmatrix} x_1 \\ x_2 \\ x_3 \end{pmatrix}$ where $y_1 = x_1^2 + x_2$ and $y_2 = 4x_2^2 + x_3^3$. Find the Jacobian of the transformation.

2. Given $f(x,y,z) = \sin(xy + y^2z)$ determine the Gradient matrix and the Hessian matrix.

3. If $f(X) = 4x_1^2 + 6x_1x_2^3$ and $X = \begin{pmatrix} x_1 \\ x_2 \end{pmatrix}$, find the gradient matrix G.

4. Find the matrix differential $ln(|X|)$.

5. Find the matrix differential of X^T.

6. If Y_i, $1 \le i \le n$ are independent identically distributed random variables with joint likelihood function

$$p(y_i|\ x_i, a_0, a_1) = \Pi_{i=1}^n \frac{y_i - (a_0 + a_1x_i)}{24} e^{-\frac{(y_i - (a_0 + a_1x_i))^2}{2}},$$

   a. Determine $LL = log(p(y|\ x_1, x_2)$.

   b. Find $\dfrac{\partial LL}{\partial a_0}$.

   c. Find $\dfrac{\partial LL}{\partial a_1}$.

7. If $f(X) = -Y log(XW)$ where X is a 2×2 matrix and $W = \begin{pmatrix} a_0 \\ a_1 \end{pmatrix}$, determine

$$\frac{\partial f}{\partial W}.$$

8. SSE is a common loss function to use in gradient descent, another common loss function is the **absolute error**

$$\sum_{i=0}^n |\hat{y}_i - y_i| = \sum_{i=0}^n |(a_0 + a_1x_i) - y_i|.$$

   if the interpolating function is $\hat{y} = a_0 + a_1(x)$, determine

$$\frac{\partial |(a_0 + a_1x_i) - y_i|}{\partial a_0} \text{ and } \frac{\partial |(a_0 + a_1x_i) - y_i|}{\partial a_1}.$$

   (Hint: Recall that $|f(x)| = f(x)$ when $f(x) \ge 0$ and $|f(x)| = -f(x)$ when $f(x) < 0$.)

9. Given the interpolation points $\{(-1, 3), (0, -4), (1, 5), (2, -6)\}$ if the loss function, $L(x)$, is the absolute error from Exercise 8, and the goal is to find a linear interpolating function $\hat{y} = a_0 + a_1 x$,

   a. Complete the following table for initial guess $a_0 = 5, a_1 = -.5$.

y_i	$\hat{y}_i = a_0 + a_1 x_i$	$\dfrac{\partial L}{\partial a_0}$	$\dfrac{\partial L}{\partial a_1}$
3	5.5		-1
-4	5	1	
5	4.5		
2	4	1	-6
Total			

   b. If the learning rate is $r = .03$, use the table from part a. to determine a new value for a_0 and a_1.

10. In Exercise 9, what happens if the initial guess is $a_0 = 0$, $a_1 = 0$.

11. Table 4.3 shows the number of COVID-19 total cases in NC over a six month period in 2020.

   a. Create a quadratic interpolant $\hat{f}(x) = a_0 + a_1 x + a_2 x^2$ with initial guesses for the coefficients $a_0 = 40000$, $a_1 = 30000$, $a_2 = 4000$ and determine the SSE for this model.

   b. Determine the gradients $\dfrac{\partial SSE}{\partial a_i}$, $0 \leq i \leq 2$ from the model in part a.

   c. Use the gradients from part b. and a learning rate $r = .0003$ to create a quadratic interpolant and calculate the SSE for the new model.

12. Given interpolating points $\{(0, 3), (1, 2), (2, -1), (3, 5)\}$, find a linear maximum likelihood interpolating function.

13. Example 5.19 uses the Softmax activation function to determine a model for COVID vaccine effectiveness probabilities. Let

$$X = \begin{pmatrix} 1 & 0 \\ 1 & 1 \\ 1 & 2 \end{pmatrix} \text{ and } Y = \begin{pmatrix} 0 \\ .3 \\ .9 \end{pmatrix}$$

the effective probabilities for 0, 1, or 2 COVID vaccines.

   a. Use a logistic sigmoid activation function with weights $W^{(0)} = \begin{pmatrix} .1 \\ .9 \end{pmatrix}$ to determine estimated values for Y.

b. Find SSE for your model in part a.

c. Update your weights from part a. using one iteration of gradient descent and a learning rate of $r = .03$. Determine the SSE with your new weights.

d. Use a ReLU activation function with weights weights $W^{(0)} = \begin{pmatrix} .1 \\ .9 \end{pmatrix}$ to determine estimated values for Y.

e. Find SSE for your model in part d.

f. Update your weights from part d. using one iteration of gradient descent and a learning rate of $r = .01$.

14. Example 5.25 uses a Gaussian activation function to estimate categories for facial recognition. Use the data given in Example 5.25, $W^{(0)} = \{.1, .9\}$ and the activation function

$$\phi(X) = \frac{w_1}{(X - \mu_1)^T(X - \mu_1)} + \frac{w_2}{(X - \mu_2)^T(X - \mu_2)}$$

with the SSE cost function to determine which category most resembles the images represented by f_1 and f_2.

6

Decision Trees and Random Forests

While previous chapters focus on techniques for making a single decision or model, this chapter focuses on ways to make a series of decisions that can incorporate previous techniques in order to best categorize data.

6.1 Decision Trees

A **decision tree** is a recursive partitioning of choices represented by a tree graph. A decision tree is similar to a flow chart and can be applied in a variety of venues. The general form of a decision tree can be seen in Figure 6.1.

Let's begin with an example of a decision tree applied to a logic statement.

Example 6.1. *Perhaps you wish to find the outcome of the logic statement $A \cap (B \cup C)$. Figure 6.2 shows a decision tree for this logic statement.*

Notice that in Figure 6.2 that two leaves are True and two leaves are False and thus both True and False are equally likely outcomes.

Typically decision trees are used for classification tasks and regression analysis. A **classification tree** is a decision tree where each node, or vertex, has a binary decision connected to it based on whether an attribute is present or not. The top node, also called the **root**, in a classification tree represents the presence of all possible attributes, and each **child** node that is a subdivision of the node above it, also called the **parent** node.

A child node may be a successor decision node or a final outcome, called the **leaf** node.

Example 6.2. *In Example 3.8, four Cinderella tales are presented with five attributes. Example 3.8 uses PCA to explore a measure of closeness between these tales. Perhaps instead, one wishes to create a decision tree to determine how close these tales are to the Disney version of the tale, based on Charles Perrault's Cendrillon.*

In addition to the data in Table 3.1, the attributes of the Cendrillion tale include

DOI: 10.1201/9781003025672-6

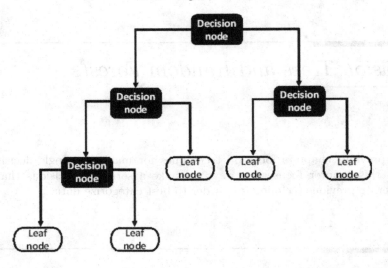

FIGURE 6.1
General Form of a Decision Tree.

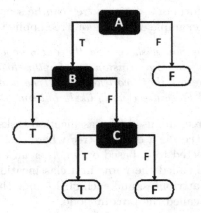

FIGURE 6.2
Example of a decision tree for logic statement $A \cap (B \cup C)$.

TABLE 6.1
Attributes for the Cendrillion table.

The heroine's name means ash	The heroine picks lentils	Magical animals are present	Action takes place in a church	The heroine buries bones
1	0	1	0	0

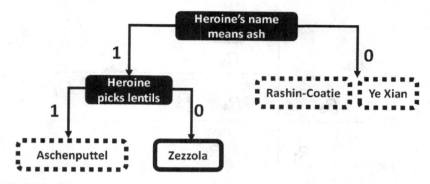

FIGURE 6.3
Decision tree for classifying Cinderella tales in Example 6.2.

If we wish to see how many attributes each of the other four tales shares (in terms of absence or presence) with Perrault's Cinderella tale, we may create a decision tree similar to the one shown in Figure 6.3.

A decision tree can also be used to divide quantitative data into regions or clusters.

Example 6.3. *In Figure 6.4, you can see that there are some clear splits in classification when $x < .05$ and $.1 < y < .15$.*

But not all decisions are created equally. Figures 6.2 and 6.5 show two different decision trees for $A \cap (B \cup C)$, which one do you prefer?
Given the choice, it is best to choose a root node to split a decision in a way that provides more clarity after the split.

A decision tree defined over a domain of categorical attributes can be converted into a discrete numerical function. Many times, data scientists will wish to determine such functions that establish criteria to measure the closeness of the data to the desired outcome. There are several standard criteria that are used.

In Section 5.6, cross-entropy was used in order to make decisions about handwritten digits. **Entropy** and the **Gini index** are some examples of splitting functions.

Entropy

If Y is a random event, then the **uncertainty** of $Y = y$ occurring can be measured by

$$log_2 \left(\frac{1}{P(Y = y)} \right) = -log_2(P(Y = y)).$$

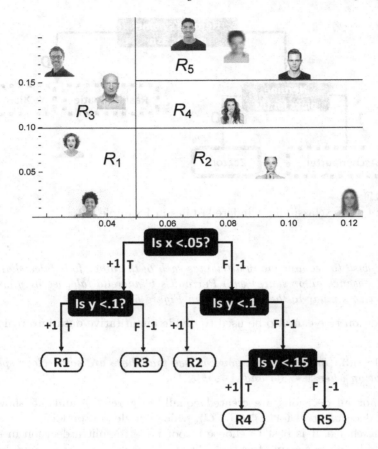

FIGURE 6.4
Example of data with splits and decision tree.

Example 6.4. *If* $Y \sim N(0,1)$ *then*

$$-log_2(P(Y=y)) = -log_2\left(\frac{1}{2\pi}e^{-y^2/2}\right).$$

Notice that when values close to the mean, $\mu = 0$, *occur, the uncertainty is close to 0 while values further away from the mean have greater uncertainty of occurring.*

The **entropy** of a random event Y is the expected uncertainty. That is, if Y is a discrete random variable, then

$$Entropy(Y) = E(-log_2(P(Y=y))) = -\sum_{y} P(Y=y)log_2(P(Y=y))$$

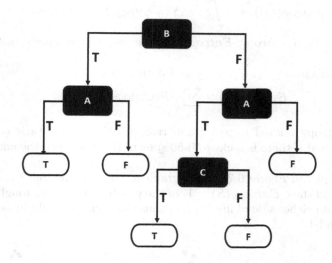

FIGURE 6.5
Another decision tree for logic statement $A \cap (B \cup C)$.

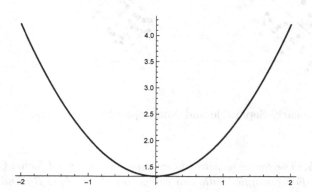

FIGURE 6.6
Uncertainty of a random event $Y \sim N(0,1)$.

and if Y is a continuous random variable,

$$Entropy(Y) = -\int_{-\infty}^{\infty} P(Y = y)log_2(P(Y = y))\ dy.$$

For a dataset S, entropy, $Entropy(S)$, is a measure of homogeneity within the dataset.

If there are classes $C_1, C_2, \ldots, C_n$ of data in S then

$$Entropy(S) = \sum_{C_i} -P(C_i)log_2(P(C_i)).$$

If the entropy related to a decision tree is 0 then the data is completely homogenous and if there is a clear 50-50 split in the data then the entropy will be equal to 1.

For example, in Figure 6.7, in the graph to the left, the line L_1, splits the two classes and thus, $Entropy(S) = 1$. In the graph on the right, roughly half of each class is on either side of line L_1 and thus the entropy would lie somewhere between 0 and 1.

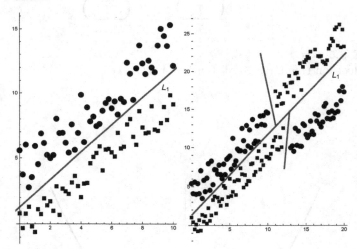

FIGURE 6.7
Example of Linearly Separable and Non-Separable Data Sets.

Example 6.5. *The facial recognition example in Figure 6.4 focuses on a variety of categories. For example, with two categories* $C_1 = R_1 \cup R_2$ *and* $C_2 = R_3 \cup R_4 \cup R_5$,

$$Entropy(C_1) = -P(C_1)log_2(P(C_1)) = -\frac{4}{10}log_2\left(\frac{4}{10}\right) \approx 0.53$$

$$Entropy(C_2) = -P(C_2)log_2(P(C_2)) = -\frac{6}{10}log_2\left(\frac{6}{10}\right) \approx 0.44$$

Thus, the entropy if using these two categories is approximately 0.97. Noting that this value is close to 1 tells us that the data is very close to a 50-50 split within the categories.

Similarly, if we choose to separate the data with categories, $C_1 = R_1$ and $C_2 = R_2 \cup R_3 \cup R_4 \cup R_5$,

$$Entropy(C_1) = -\frac{2}{10} log_2 \left(\frac{2}{10} \right) \approx 0.46$$

$$Entropy(C_2) = -\frac{8}{10} log_2 \left(\frac{8}{10} \right) \approx 0.26$$

with a total entropy of .73 the data is less of an even split between categories.

Another indicator of strong splits in decisions is the Gini Impurity measure.

Gini Impurity

The idea behind the Gini Impurity measure is to determine the probability that the nodes are chosen in a way that a random sample of data is classified incorrectly.

$$Gini\ Impurity = \sum_{i=1}^{n} p_i(1 - p_i)$$

$$= 1 - \sum_{i=1}^{n} p_i^2$$

where $P = (p_1, p_2, \ldots, p_n)$ and p_i is the probability that an object is classified in class $i = 1, 2, \ldots, n$.

For example, assume we randomly select 10 data points as in Figure 6.8, with two categories. If a decision criteria in Node 1 creates a split shown by Nodes 2 and 3, we can then see that the

$$\text{Gini Impurity of Node 2} = 1 - \frac{3}{4}^2 - \frac{1}{4}^2 = \frac{3}{8}$$

$$\text{whereas the Gini Impurity of Node 3} = 1 - \frac{1}{3}^2 - \frac{2}{3}^2 = \frac{4}{9}.$$

The Gini Impurity ranges from 0 to 1 and the higher the value the more impure the nodes of the decision tree are. This measure is only used for measuring categorical data. Other measures must be used for quantitative or continuous data.

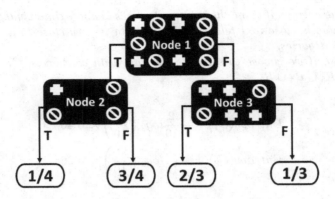

FIGURE 6.8
Example of decision criteria splitting.

Example 6.6. *Revisiting the facial recognition problem in Figure 6.4 with categories $C_1 = R_1 \cup R_2$ and $C_2 = R_3 \cup R_4 \cup R_5$,*

$$Gini\ Impurity = 1 - (P(C_1))^2 - (P(C_2))^2 = 1 - \left(\frac{4}{10}\right)^2 - \left(\frac{6}{10}\right)^2 = \frac{12}{25}$$

while categories defined as $C_1 = R_1$ and $C_2 = R_2 \cup R_3 \cup R_4 \cup R_5$ give a

$$Gini\ Impurity = 1 - \left(\frac{1}{10}\right)^2 - \left(\frac{8}{10}\right)^2 = \frac{7}{20}$$

Example 6.7. *For the decision tree presented in Figure 6.2, we calculate the Gini index for nodes A, B, and C where $p = (1 - p) = \frac{1}{2}$ for each of the nodes and thus*

$$Gini\ impurity\ A = Gini\ impurity\ B = Gini\ impurity\ C = \frac{1}{2},$$

where the decision tree in Figure 6.5, for the same logistic statement, has a Gini impurity $= \frac{13}{25}$.

Another way to think about the Gini impurity measure is that if P is a probability matrix where $P_{i,j} = p_i$ when $i = j$ and $P_{i,j} = 0$ otherwise. Then

$$Gini\ impurity = Tr(I - P^2).$$

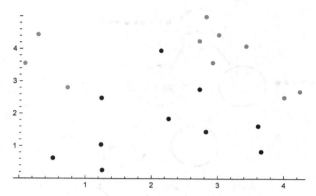

FIGURE 6.9
Training set for decision tree categorization example.

6.1.1 Decision Trees Regression

Typical linear regression models work to linearly separate data, whereas decision tree models work to separate data that is not linearly separable. A visualization of a linearly separable set of data versus one which is not linearly separable can be seen in Figure 6.7.

In this section, we will focus on (1) Decision Tree Linear Regression as a categorization problem and (2) Decision Trees as a regression model. Later, we will take a brief look at how one might combine these ideas with those explored in Chapter 4 to create even better models.

Since the data that will be the focus of decision tree linear regression is typically linearly non-separable, the mathematical algorithm discussed is recursive in nature, starting at the root of the tree.

When using decision trees to determine clustering or linear regression models, it is important to hold back a set of data for a test data set for learning purposes. We will follow the steps of this recursive algorithm with an example related to the data in Figure 6.9.

We begin with a trained decision tree with a root node containing our entire test data set and a splitting feature.

With a root node as seen in Figure 6.10, notice that we begin by including all of the data points. Under the root condition, in this case $x_1 > 2$, child nodes are included based on whether the root condition is True or False.

After this first decision, the right child node only has data from a single category. We call this node a **pure node**. Notice that the first child node on the right in Figure 6.9 is not a pure node.

If we are given some data not in this training set, for example, (2, 1) and (0.5, 3), then we can follow the decision tree to decide which category to place these data points.

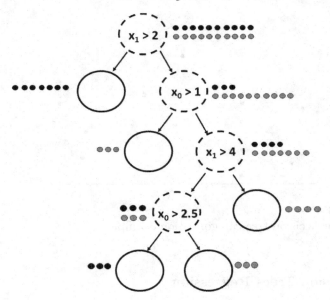

FIGURE 6.10

Example decision tree for categorization example, using training data in Figure 6.9.

Since all of the child nodes are pure, $(2, 1) \to \bullet$ and $(0.5, 3) \to \circ$.

In this example, there were many choices for conditions to split the training data, we discuss now how we best choose these conditions.

One way to do so is with a **greedy algorithm** in which conditions are considered by brute force. Unfortunately, this algorithm may still not determine the best splitting as once a condition on a root is determined the algorithm does not go back up the tree.

The steps for the greedy algorithm include:

- Setting a root condition that produces a Gini impurity close to 1 (In the above example, for $x_1 > 2$, the Gini impurity is 1)

- For each possible splitting rule, calculate IG,

$$Information\ Gain = Entropy(parent) - \sum_i w_i Entropy(child).$$

For the data shown in Figure 6.11, with root condition $x_1 > 2$, the child conditions are shown in Table 6.2.

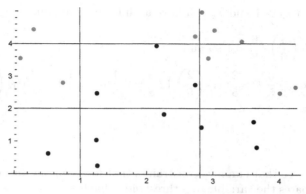

FIGURE 6.11
Example of decision splitting for categorization example, using training data in
Figure 6.9.

TABLE 6.2
Example initial splits and corresponding entropy and information gained
based on training data in Figure 6.9.

Condition	Split, S_1,S_2	Entropy for S_1	Entropy for S_2	IG
$x_0 > 1$	(3,0), (7,3)	0	$-.3log_2(.3) - .7log_2(.7)$ ≈ 0.88	$1 - \frac{3}{13} \cdot 0$ $-\frac{10}{13} \cdot 0.88 \approx 0.32$
$x_0 > 2$	(3,1), (7,2)	0.81	0.76	0.225
$x_0 > 3$	(6,3), (4,0)	0.92	0	0.36
$x_0 > 4$	(8,3), (2,0)	0.85	0	0.28
$x_1 > 3$	(3,2), (7,1)	0.97	0.54	0.295
$x_1 > 4$	(5,3), (5,0)	0.95	0	0.41

• Choose a new condition that maximizes the information gain, IG.

A pure child will produce an entropy of 0. Ideally, we would choose a smaller
grid and larger variety of conditions; however, with the choices presented in
Table 6.2, three pure child nodes exists.

The best choice for the next condition from those presented in Table 6.2 is
$x_1 > 4$.

• Repeat Steps 1-3.

With a new parent node and condition $x_1 > 4$, focus on the region

$$0 \leq x_0 \leq 4, 2 \leq x_1 \leq 4.$$

The entropy of this new parent node is 0.95.

The condition $x_0 > 1$ and $x_1 > 3$ have an information gain of

$$0.95 - log_2\left(\frac{3}{6}\right) \cdot \frac{6}{8} = 1.95 \text{ and}$$

$$.95 - \left(\frac{3}{5}log_2\left(\frac{3}{5}\right) + \frac{2}{5}log_2\left(\frac{2}{5}\right)\right) \cdot \frac{5}{8} - \left(\frac{1}{3}log_2\left(\frac{1}{3}\right) + \frac{2}{3}log_2\left(\frac{2}{3}\right)\right) \cdot \frac{3}{8}$$

$$= 2.13078$$

respectively.

A decision tree that employs constraints that focus on one variable, x_i, at a time and compares the variable to a threshold value b, such as $x_0 > 4$ or $x_1 > 2$, is called a **univariate decision tree**.

A decision tree that is developed from constraints that are linear combination of the variables, or attributes, and compares to a threshold value b, such as $2x_0 + 4x_1 > 6$ and $\frac{7}{4}x_0 - x_1 > 8$, is called an **oblique decision tree**.

An oblique decision tree has constraints of the form

$$Ax > b.$$

Example 6.8. *An oblique decision tree associated with the data in Figure 6.10, may start with the root constraint $x_1 > 2$ and then follow with child constraint*

$$x_1 + \frac{32}{19}x_0 > 8,$$

seen in Figure 6.12. Compared to the results presented previously, with this new root condition,

$$Entropy = -log_2\left(\frac{1}{2}\right) = 1$$

with an information gain,

$$IG = 1 - 0 \cdot \frac{7}{13} - log_2\left(\frac{1}{2}\right) \cdot \frac{6}{13} = \frac{19}{13}.$$

An additional child constraint of $x_0 - 2x_1 > \frac{1}{2}$ would yield an additional pure decision.

Thus far, our examples of decision tree algorithms focus on categorization or classification. Similar techniques can be used for regression purposes as well. When using these algorithms for regression purposes, we can envision the regression problem as a classification problem, or a combination of decision trees and linear regression techniques can be employed.

Example 6.9 shows how a regression problem can be transformed into a classification problem in order to use decision tree algorithms.

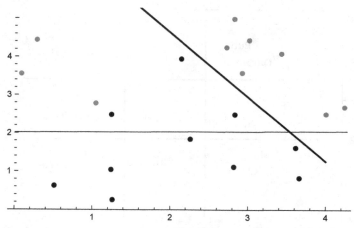

FIGURE 6.12
Visualization of an example decision in an oblique decision tree related to data.

Example 6.9. *Case Study 4.9 looked at using interpolation techniques to explore success for black students in higher education, using data listed in Table 6.3. For this problem, we will approach the goal of estimating graduation rates of black students, based on the tuition of an institution and the fraction of the student body which consists of black students, with decision tree algorithms.*

TABLE 6.3
Sample of college data to predict graduation rates of black students in higher education institutions.

	College	InState Tuition	Total Enrollment	Total Minority	Graduation Rate for Black Students in 2020
1	Carnegie Mellon	69883	12587	3392	80
2	Clemson	25802	21857	2849	63.4
3	Florida State	17332	41226	12238	72.5
4	James Madison	22108	20855	3574	74.1
5	High Point	49248	4399	668	74.3
6	New England College	52136	2399	638	19.2
7	Old Dominion	22492	24932	10059	53.14
8	Providence College	65090	4533	652	84.8
9	Rutgers	27680	48378	21656	79.86
10	UConn	28604	26541	6255	71.05
11	Penn	71200	24806	7805	94.22
12	William and Mary	35636	8437	2135	88.18

Figure 6.13 shows the 12 schools' tuition versus graduation rate of black students for the respective schools.

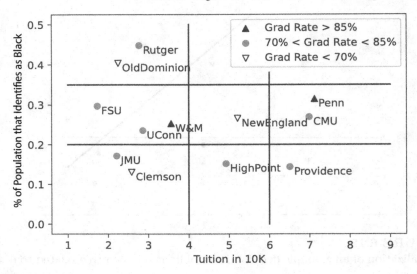

FIGURE 6.13
Tuition versus graduation rate of black students.

One can create a decision tree such as the one found in Figure 6.14 which attempts to separate universities into three categories, those with a graduation rate of less than 70%, those with a graduation rate of between 70% and 85% and those above 85%.

With this information, a university with a tuition of $75,000 and a black student population of 25% would have an estimated graduation rate, which is the average of those satisfying the same criteria.

$$\frac{92.22(U.of Penn) + 80(Carnegie Mellon)}{2} = 86.11\% \text{ graduation rate.}$$

Whereas a university with a tuition of $30,000 and a black student population of 20% would require an average graduation rate of Clemson and James Madison Universities' rates, producing an estimated graduation rate of 68.75%.

Since there are three categories in this decision regression problem, the root constraint Tuition > 60 K has an entropy of

$$-\frac{2}{12}log_2\left(\frac{2}{12}\right) - \frac{3}{12}log_2\left(\frac{3}{12}\right) - \frac{7}{12}log_2\left(\frac{7}{12}\right) \approx 1.384$$

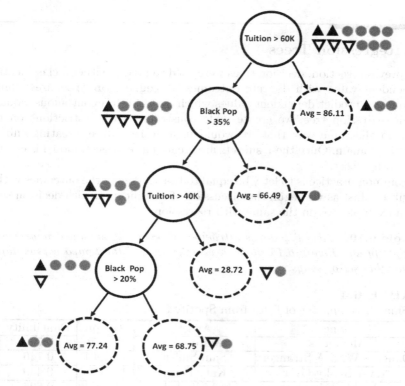

FIGURE 6.14
Example decision tree for data in Table 6.3.

and information gain from the split,

$$1.384 - \frac{9}{12}\left(-\frac{1}{9}log_2\left(\frac{1}{9}\right) - \frac{3}{9}log_2\left(\frac{3}{9}\right) - \frac{5}{9}log_2\left(\frac{5}{9}\right)\right)$$
$$+ \frac{3}{12}\left(-\frac{1}{3}log_2\left(\frac{1}{3}\right) - \frac{2}{3}log_2\left(\frac{2}{6}\right)\right) \approx .14$$

Notice that this decision does not split the data in order to produce a pure child node, so the information gain is small. Better approaches may be to consider an oblique decision tree or to use a decision tree to create regions, or subsets of data, which would then be used to create a piecewise function with more traditional regression techniques. The later will be explored further in the next section.

6.2   Regression Trees

In the previous section, decision trees were used to predict either a classification or dependent value with discrete outcomes. A **regression tree** uses similar techniques to predict dependent values which may take on continuous values.

In a regression tree, we create node constraints with restrictions on the domain in the same way that we would in a decision tree, creating subsets within the domain. Once these subsets are created a regression model is created within each subset.

A common practice is to let $\hat{y}$ be equal to the average of the outcomes within each subset. Just as we begun the discussion of classification with decision trees with an example, we do the same with regression.

Example 6.10. *Table 6.4 shows attributes of popularly streamed remix songs from Spotify and Figure 6.15 shows a visualization of the tempo versus dancibility for this set of songs.*

TABLE 6.4
A Small Training Set of Data from Spotify Streaming.

Song	Artist	Tempo	Dancibility
Memories	Maroon 5	99.972	0.726
Dancing With A Stranger	Sam Smith	111.961	0.746
Never Really Over	Katy Perry	112.648	0.449
All My Friends	AJ Mitchell	118.051	0.694
Say My Name	David Guetta	120.002	0.678
Call You Mine	The Chainsmokers	121.956	0.718

The reader can find a larger set of Spotify streaming data at GitHub link 17. Notice that although one might fit a least squares regression line to this data, it might be more beneficial to separate the data from the first part of the data set from the remaining data points.

Given the set of interpolant points $S = \{(x_1, y_1), (x_2, y_2), \ldots, (x_n, y_n)\}$, one might wish to determine a root constraint for the tree by partitioning

$$S = S_1 \cup S_2 = \{(x_1, y_1), \ldots, (x_k, y_k)\} \cup \{(x_{k+1}, y_{k+1}), \ldots, (x_n, y_n)\}.$$

One way to choose this partition is that if μ_1, μ_2 are the mean of $\{y_1, \ldots, y_k\}$ and $\{y_{k+1}, \ldots, y_n\}$ respectively and $\hat{y}_i = \mu_1$ for $1 \leq i \leq k$ and μ_2 otherwise, then the value of i is chosen to minimize some cost, such as SSE.

Figure 6.16 shows a visualization of partitioning the domain in this way along with the respective sum of squared errors. In this example, this would lead

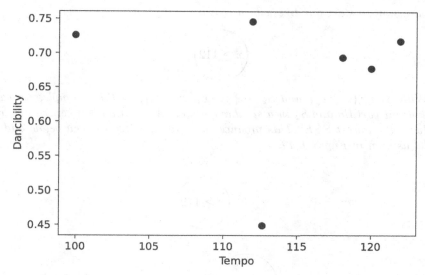

FIGURE 6.15
Spotify Streaming Training Data Set.

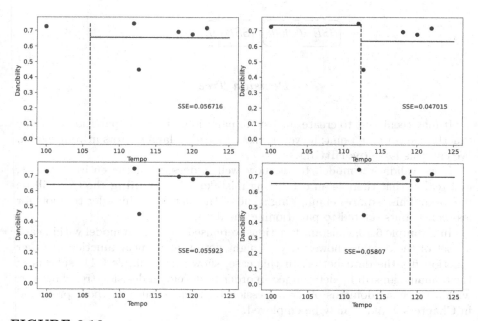

FIGURE 6.16
Determining a Constraint to Partition a Training Set.

to a decision to start our tree with a root constraint of

With $S_1 = \{x_1,\ x_2\}$ and $S_2 = \{x_3,\ x_4,\ x_5,\ x_6\}$, in this example, we can think about partitioning S_2 in a similar manner. A partition between x_3 and x_4 produces the lowest SSE. This produces a regression tree as seen below and a model as seen in Figure 6.17.

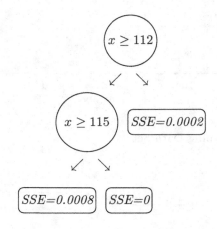

Decision Tree

It may seem best to create such small partitions that each partition contains exactly one training point; however, this can lead to larger errors in your model. We call this issue **overfitting**.

If we validate a model, to see how well it does, with the entire data set, and realize that there is an overfitting mistake, then **pruning** the tree will be required. This requires eliminating some of the leaf layers in order to allow for less constraints related to partitioning the data.

In Example 6.10, constant functions were used to create a model within each subset of the domain; however, you might note that a linear function or other function fits the data better. In this case, shown in Example 6.11, subsets of the domain can still be determined in order to develop a decision tree; however, within these subdomains other regression techniques, such as those presented in Chapters 4 and 5, may be employed.

Example 6.11. *Table 6.5 shows a small training set of U.S. poverty data throughout time, and Figure 6.18 shows one way in which you could choose to partition the data in order to create a piecewise model.*

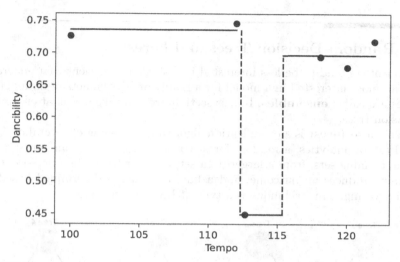

FIGURE 6.17
A Regression Tree Model.

TABLE 6.5
U.S. Poverty Rate over Time (Training Data).

2001	11.7	2013	14.8
2005	12.6	2015	14.8
2008	13.2	2018	11.8
2010	15.1	2020	11.4

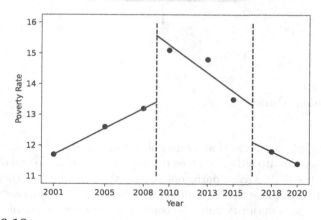

FIGURE 6.18
An Example of a Regression Tree Model for Poverty Rates in the U.S. (2001–2020) [32].

6.3   Random Decision Trees and Forests

Data scientists tend to be less interested in a single experiment that generates data and more interested in a model for a family of experiments of data. These families are called **ensembles**. In this section, we will take a look at ensembles of decision trees.

A **random forest** is a classification algorithm consisting of several decision trees. In data analytics, a random forest consists of choosing random samples of data, training sets, from a larger data set, in order to create a decision tree, each tree produces an outcome, and a final outcome is determined based on majority voting. This technique is a type of **bagging** technique.

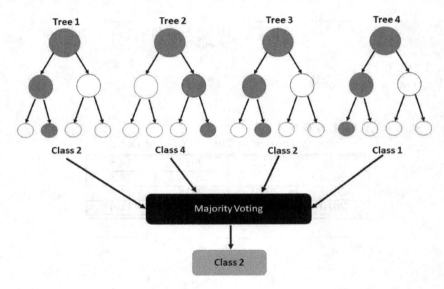

FIGURE 6.19
Example Random Forest.

So for example let's say that you choose four random training sets, with replacement, and create a decision tree to classify the data into one of four class as seen in Figure 6.19. Given data, not within the training set, these decision trees can be used to place the data into classes, where the final classification can be made based on a majority rules decision. An additional example of random forest implementation can be see in the Case Study in Section 6.5.

6.4 CASE STUDY: Entropy of Wordle

The goal of Wordle is to come up with a secret five letter word within six guesses. Whether your game of the day is Wordle, Quordle, or even Octordle, you may have a go to word or set of words as part of a personal strategy to find the word or words of the day. You might think to use many vowels and start your game off with the initial guess "adieu", or introduce common letters and try an initial guess like "snort", or finally you might be a risk taker and go for a word that has uncommon letters that you would like to eliminate like "zesty".

Although the actual Wordle game uses colored green, yellow, and gray squares to represent particular qualities to the guess, the following colored squares will be used in this analysis

$green \rightarrow$ ☐ $\rightarrow$ guess letter present in the correct location

$yellow \rightarrow$ ▦ $\rightarrow$ guess letter present in the incorrect location

$gray \rightarrow$ ☐ $\rightarrow$ guess letter not present

There are roughly 13,000 five letter words that are acceptable in the Wordle game and about 15900 five-letter words found in the word bank at GitHub link 41.

So how can we use entropy to choose our words wisely? A player that chooses a word like **ZESTY** for the first guess and gets the Z correct and in the correct location reduces the word choice list to 1200. According to the Scrabble dictionary, roughly 9% of five-letter words start with the letter Z.

Thus if we chose to make a decision purely based on whether the first letter is a Z, the information gained by playing $ZESTY$ with this outcome is

$$IG = 1 - .09 \cdot log_2(.09) - .91 \cdot log_2(.91) \approx 1.44.$$

However, if one wishes to make a decision based on whether the Y was correct, it is important to note that there are 1254 five-letter words ending with Y. A decision made based on a correctly placed Y at the end of the word provides an

$$IG = 1 - \left(\frac{1254}{13000}\right) \cdot log_2\left(\frac{1254}{13000}\right) - \left(\frac{11746}{13000}\right) \cdot log_2\left(\frac{11746}{13000}\right) \approx 1.19.$$

Notice that the larger the probability of an event occurring, the smaller the information gain if the event occurs. In order to pick the best guess, it is helpful to choose the word with the smallest entropy and thus largest information gained. In this case study, we show a simplified version of what one might do.

From a beginning guess word, such as $ZESTY$, one might wish to find the probability of each outcome occurring. Figure 6.20 shows the probability of a word ending with a certain letter. Notice that a five-letter word with the

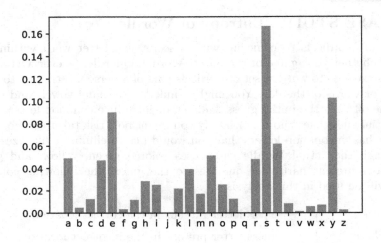

FIGURE 6.20
Probability of a five-letter word ending in a certain letter.

S occurring in the last spot is most likely and thus would produce the least information gained, while one ending in B or Z would provide a large information gain.

If you did not want to focus on just one letter of your guess when making decisions, you could look at the probability of outcomes occurring, such as in Figure 6.21. The overall distribution of outcomes for the guesses *ZESTY* and *TARES* can be seen in Figure 6.22.

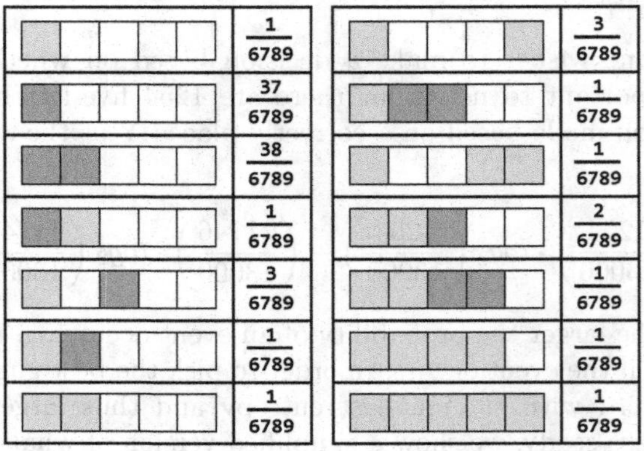

FIGURE 6.21
Probability of specific outcomes occurring with a guess of *ZESTY*.

Notice that the probability distribution for the guess *ZESTY* is taller and thinner than that for the guess *TARES*. The probability of an event occurring and the entropy associated with the event are inversely proportional. Thus the closer the distribution is to uniform, flat, the more information is provided by the guess.

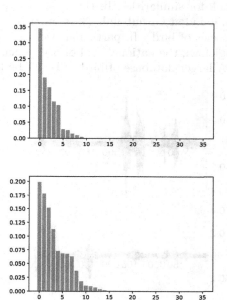

FIGURE 6.22
Probability Distribution for Outcomes Given the Guess *ZESTY* (TOP) and *TARES* (BOTTOM).

We use the heights of the bars in the simplified probability distribution shown in Figure 6.22 as the probabilities for the entropy of the guess. Using the simplified distribution from Python code found at GitHub link 42, the entropy, or the expected information value, for the guesses *ZESTY* and *TARES* are approximately 2.6 and 3.24 respectively.

So if you did start your game with a guess of *ZESTY* and got an outcome of

what would your next guess be?
zapas, zapus, zarfs, ziffs, zills, zincs, zings, zoism, zooks, zooms, zoons, zoris, zulus, zunis

Python and R code for this case study can be found at Github link 37 and 39 respectively.

6.5   CASE STUDY: Bird Call Identification

Many times, species identification techniques require capture and release. A less invasive species identification technique is to record the calls of the species, in this case birds, and look for similarities. In this case study, we are given data from several bird calls, found at Github link 43, in order to develop a model for the classification of species of birds. In particular, we wish to identify the call of the Sri Lankan Junglefowl, the national bird of Sri Lanka, among other bird calls from Sri Lanka. A larger database of bird calls can be found at [43].

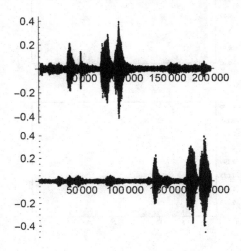

FIGURE 6.23
Visualization of soundwaves of two different Sri Lankan Junglefowl.

It is important to note that, in reality, this type of data could consist of data blocks from very different times in the birds calling cycle. Therefore, one bird's call may just be a translated version of another's. Additionally, certain birds could be closer or calling louder, and thus the scale on the call may not be significant in the identification of the species.

What might be more important to observe is the dominant frequencies present in the call. In this case study, we discuss how to transform data into a frequency domain for this type of analysis.

Given a discrete data set $\{y_1, y_2, \ldots, y_n\}$, the **Discrete Fourier Transform** is defined as

$$F_k = \sum_{j=1}^{n-1} y_j e^{\frac{-2\pi i}{n} kj}.$$

Figure 6.24 shows the frequency distribution for several Sri Lankan jungle-fowl calls. For more information about Fourier analysis see [28].

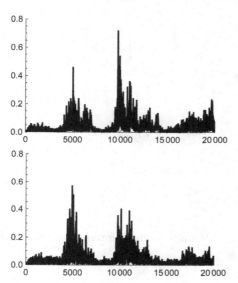

FIGURE 6.24
Visualization of Frequencies in Sri Lanka junglefowl calls from Figure 6.23.

From this, one might wish to study what the peak frequencies are, or the width between peak frequencies, among other characteristics.

If p_1 and p_2 are the two frequencies, 1K Hz as the units, occurring the most in the call, Figure 6.25 shows p_1 versus the absolute relative increase between p_1 and p_2,

$$\left| \frac{p_1 - p_2}{p_1} \right|.$$

The dashed lines in Figure 6.25 might be considered as criteria for a decision tree. Given this information, a bird call with a peak frequency of 20K Hz and an absolute relative increase of 9Hz <u>WOULD</u> be identify as a Sri Lankan junglefowl while a bird call with a peak frequency of 25K Hz and an absolute relative increase of 30Hz <u>WOULD NOT</u> be identify as a Sri Lankan junglefowl.

Github links 38 and 40 provide Python and R code for this case study.

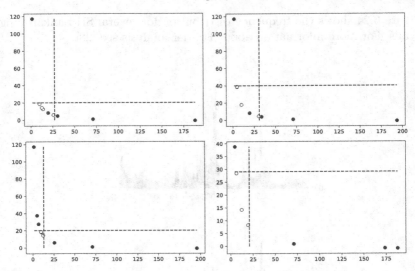

FIGURE 6.25
Visualization of four outcomes from different training data sets, jungle fowl
denoted by ○.

6.6 Exercises

1. Given the data in Figure 6.26 and root condition, $x > 0$,

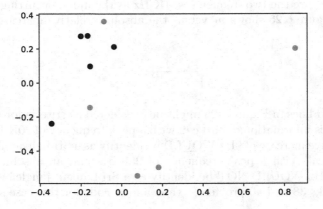

FIGURE 6.26
Data for Exercise 1.

a. Find the entropy and information gained for the root criteria $x > 0$.

b. Determine a root condition that will create a pure child node.

c. Create a decision tree which maximizes information gain at each stage with the root node condition $x > 0$.

2. Spotify training data in Table 6.6 shows loudness and acoustic factors for several songs. Use techniques presented in Example 6.10 to determine a regression tree.

TABLE 6.6

Spotify training data.

Loudness	Acousticness
-8.078	0.000464
-4.234	0.00105
-9.533	0.0612
-6.36	0.112
-7.734	0.0553
-7.165	0.379

3. Table 6.7 shows a training data set from a disease database based on burden by cause [41].

TABLE 6.7

Training data set from a disease database.

Substance Abuse	Interpersonal Violence	Self-harm
1.17	2.29	0.56
1.29	1.29	0.36
0.48	1.81	2.98
3.52	3.52	1.52
0.26	0.16	0.35
0.24	1.80	1.46
0.60	0.22	0.38
0.68	0.22	0.38
2.43	1.15	0.63
1.37	0.31	0.42
0.73	0.24	0.41
0.58	3.29	2.16

a. Create a regression tree to model interpersonal violence versus self-harm.

b. If the self harm measure is placed into 3 categories, Category 1 (low, less than .5), Category 2 (medium, between .5 and 1), and Category 2 (high, above 2), create a graph of substance abuse

Country	Year	Self harm	Substance abuse disorders	Interpersonal violence
Sri Lanka	2019	3.47	1.091	0.87

    versus interpersonal violence that also shows the category of self harm.

c. Use the graph in part b. to create a decision tree to categorize the self harm measure.

d. Use the decision tree in part c. to predict the category of self harm for Sri Lanka in 2019.

e. Discuss why a random forest model may produce better results than the single decision tree model from part c.

f. Using the dataset found at Github link 50, chose at least 5 random training sets. For each random training set, replicate steps b. through d. From this random forest algorithm again predict the category of self harm for Sri Lanka in 2019.

7

Random Matrices and Covariance Estimate

7.1 Introduction to Random Matrices

So far, we have used matrices with fixed values to model scenarios in data science. Here we will explore the idea of matrices whose entries are determined at random from a probability density function. Such a matrix is called a **random matrix** and each entry is consider to be a random variable.

For example, we can define a 3×3 matrix, X, with entries that are randomly distributed with a standard normal distribution, $N(0,1)$. One example of a random matrix, X, such that $X_{i,j} \sim N(0,1)$ is

$$X = \begin{pmatrix} 0.290226 & 0.91635 & 0.0998479 \\ 0.882873 & -0.615393 & -0.902417 \\ -0.0974429 & -0.855375 & -0.665516 \end{pmatrix}.$$

The matrix X is not symmetric; however, matrices such as XX^T and $X + X^T$ will always be symmetric matrices with real eigenvalues.

A $n \times n$ matrix, M, with entries $m_{i,j}$ is a called a **Wigner matrix** if the entries of M, $m_{i,j}$ for $1 \leq i \leq j \leq n$, are randomly i.i.d.

In this section, we will focus on random matrices whose entries come from a Gaussian distribution,

$$X = \begin{pmatrix} X_{1,1} & X_{1,2} & X_{1,3} & \dots & X_{1,n} \\ X_{2,1} & X_{2,2} & X_{2,3} & \dots & X_{2,n} \\ X_{3,1} & X_{3,2} & X_{3,3} & \dots & X_{3,n} \\ \vdots & \ddots & \ddots & \ddots & \vdots \\ X_{m,1} & X_{m,2} & X_{m,3} & \dots & X_{m,n} \end{pmatrix},$$

where $X_{i,j} \sim N(\mu,\sigma^2)$ are i.i.d, $1 \leq i \leq m$, $1 \leq j \leq n$.

Typically, random matrices of importance in application fall into three general categories based on whether their entries are real, complex, or quaternionic random variables. In this chapter, we will be focusing on these three types of random matrices and the behavior of their eigenvalues.

DOI: 10.1201/9781003025672-7

Before taking a deep dive into these three families of random matrices, we will explore the effects of randomness on the stability of algorithms in linear algebra.

7.2   Stability

We have learned that, in linear algebra, if a matrix A is an invertible matrix then the system $Ax = b$ has a unique solution,

$$x = A^{-1}b.$$

However, when dealing with a random matrix, $A + E$, such that $E_{i,j} \sim N(\mu, \sigma^2)$, computational solutions may be incorrect or may not exist, typically due to round off error.

Example 7.1. *One can see the solution to the system*

$$\begin{pmatrix} -0.1 & -1 \\ 0 & 1 \end{pmatrix} \begin{pmatrix} x_1 \\ x_2 \end{pmatrix} = \begin{pmatrix} 0.1 \\ -1 \end{pmatrix},$$

$$\begin{pmatrix} x_1 \\ x_2 \end{pmatrix} = \begin{pmatrix} 0 \\ -1 \end{pmatrix}.$$

However if $A = \begin{pmatrix} -0.01 & -1 \\ 0 & 1 \end{pmatrix}$, $E_{i,j} \sim N(0, 0.01)$, *and*

$$A + E = \begin{pmatrix} 0 & -1 \\ -0.02 & 1.02 \end{pmatrix},$$

then

$$\begin{pmatrix} x_1 \\ x_2 \end{pmatrix} = \begin{pmatrix} -1 \\ -1 \end{pmatrix}.$$

The **condition number** of a matrix, A, $\kappa(A)$, is a number that represents how fast a solution, x, to the linear system $Ax = b$ is changing with changes in b.

$$\kappa(A) = \begin{cases} ||A|| \cdot ||A^{-1}||, & \text{if } A \text{ is invertible,} \\ \infty, & \text{otherwise.} \end{cases}$$

Example 7.2.

$$\kappa \left(\begin{pmatrix} -0.1 & -1 \\ 0 & 1 \end{pmatrix} \right) = \left\| \begin{pmatrix} -0.1 & -1 \\ 0 & 1 \end{pmatrix} \right\| \cdot \left\| \begin{pmatrix} -0.1 & -1 \\ 0 & 1 \end{pmatrix}^{-1} \right\|$$

$$= \left\| \begin{pmatrix} -0.1 & -1 \\ 0 & 1 \end{pmatrix} \right\| \cdot \left\| \begin{pmatrix} -10 & -10 \\ 0 & 1 \end{pmatrix} \right\|$$

$$\approx 1.41598 \cdot 14.1598 = 20.05$$

It is interesting to note that the closer a matrix, A, is to singular, that is the closer the determinant of the matrix, $|A|$, is to 0, the larger the condition number, implying the more unstable solutions are to $Ax = b$.

Example 7.3. *For random 2×2 matrix*

$$A = \begin{pmatrix} 0.93426714 & -1.3138943 \\ 0.10796494 & 0.07159739 \end{pmatrix},$$

$\kappa(A) \approx 12.451$, *whereas a defined matrix without random entries such as*

$$B = \begin{pmatrix} 1 & 0 \\ 1 & 1 \end{pmatrix}$$

has a condition number, $\kappa(B) \approx 2.618$.

When dealing with a matrix with a higher condition number, the accuracy of results will be lower in solving problems related to the matrix. This includes the stability of the eigensystem of the matrix.

Theorem 14. *(Bauer-Fike Theorem) If A is an $n \times n$ diagonalizable matrix with complex entries such that*

$$A = VDV^{-1},$$

where the columns of V are eigenvectors of A, D is a diagonal matrix of corresponding eigenvalues and E is an $n \times n$ matrix used to perturb the entries of A slightly, then if λ_{A+E} is an eigenvalue of $A + E$, there exists an eigenvalue of A, λ_A, such that

$$|\lambda_{A+E} - \lambda_A| \le \kappa(V)\|E\|.$$

Example 7.4. *If*

$$A = \begin{pmatrix} -0.01 & 1 \\ 0 & 1 \end{pmatrix}.$$

With the eigenvectors of A as the columns of V,

$$V = \begin{pmatrix} 1 & -\frac{1}{\sqrt{10}} \\ 0 & \frac{3}{\sqrt{10}} \end{pmatrix}$$

$$A = V \begin{pmatrix} 4 & 0 \\ 0 & 1 \end{pmatrix} V^{-1}.$$

Since the columns of matrix V are eigenvectors, they are not unique. We will explore the matrix V with unit eigenvectors as the columns.

$$\kappa(V) = ||V|| \cdot ||V^{-1}|| \approx 1.38743.$$

If $E = \begin{pmatrix} -0.1 & 0.1 \\ -0.01 & 0.2 \end{pmatrix}$,

$$A + E = \begin{pmatrix} 3.9 & 1.1 \\ -0.01 & 1.2 \end{pmatrix},$$

$||E|| \approx 0.230935$, and $\lambda_{A+E} \approx 3.89592, 1.20408$.
Notice that

$$0.10408 = |\rho(A + E) - \rho(A)| \leq \kappa(V)||E|| = 0.320405.$$

The **pseudospectrum** of a matrix A is the family of eigenvalues of $A + E$.

Typically data scientists deal with matrices that have errors, also called **noisy matrices**. Because of this, the pseudospectrum of noisy matrices are of particular interest.

Example 7.5. *With $A = \begin{pmatrix} 1 & -2 \\ 0 & 1/2 \end{pmatrix}$ and $E_{i,j} \sim N(0,1)$, if*

$$E = \begin{pmatrix} 1 & 0 \\ 0 & 1 \end{pmatrix}$$

then the eigenvalues of $A + E$ are 2 and $\frac{3}{2}$. However, if

$$E = \begin{pmatrix} 0 & 1 \\ 1 & \frac{1}{2} \end{pmatrix},$$

then the eigenvalues of $A + E$ are $1 + i$ and $1 - i$. Figure 7.1 shows a graph representing the eigenvalues and the pseudospectrum of A.

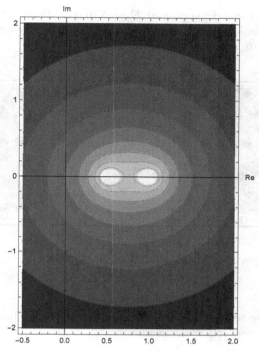

FIGURE 7.1
Visualization of a pseudospectrum.

Example 7.6. *A **Toeplitz matrix** is a band diagonal matrix of the form*

$$
T = \begin{pmatrix}
a_0 & a_{-1} & a_{-2} & \cdots & a_{-(n-1)} \\
a_1 & a_0 & a_{-1} & \ddots & \ddots \\
a_2 & a_1 & a_0 & a_{-1} & \ddots \\
\vdots & \ddots & \ddots & \ddots & a_{-1} \\
a_{(n-1)} & \cdots & a_2 & a_1 & a_0
\end{pmatrix}.
$$

If T is an upper triangular Toeplitz matrix with a single eigenvalue such as

$$
T = \begin{pmatrix}
3 & 4 & 2 & 0 \\
0 & 3 & 4 & 2 \\
0 & 0 & 3 & 4 \\
0 & 0 & 0 & 3
\end{pmatrix}
$$

and E is a random matrix such that $E \sim N(0,\sigma^2)$, Figure 7.2 shows an example of a visualization of the eigenvalues of T and perturbed matrix $T + E$, pseudospectrum, with $\sigma^2 = .01$ and $\sigma^2 = .25$.

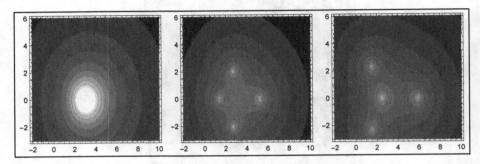

FIGURE 7.2
Visualization of a pseudospectrum, $\sigma^2 = 0, .01, .25$.

Throughout the rest of this chapter, we take a look at families of random matrices and their respective eigenvalues.

7.3   Gaussian Orthogonal Ensemble

Models for a family of matrices of data are referred to as **matrix ensembles**.

Recall from Chapter 5, that if X and Y are i.i.d. random variables where $X, Y \sim N(\mu,\sigma^2)$ then the joint distribution of X and Y,

$$f_{XY}(x,y) \sim N\left(\mu, \frac{\sigma^2}{2}\right).$$

In this section, we wish to look at the distribution of linear combinations of X and Y, such as $X + Y$ or $\frac{X}{2}$, where X and Y are random matrices, and the eigenvalues associated with them.

Theorem 15. *If $X_1, X_2, \ldots, X_n$ are i.i.d. normal random variables with means $\mu_1, \mu_2, \ldots, \mu_n$ and variances $\sigma_1^2, \sigma_2^2, \ldots, \sigma_n^2$ then the random variable described by the linear combination of $X_1, X_2, \ldots, X_n$,*

$$Z = \sum_{i=1}^{n} c_i X_i,$$

where c_i, $i = 1, 2, \ldots, n$ are scalars, has a normal distribution

$$Z \sim N\left(\sum_{i=1}^{n} c_i \mu_i, \sum_{i=1}^{n} c_i^2 \sigma_i^2\right).$$

If X is a random matrix where $X_{i,j} \sim N(0, 1)$, then

$$H = \frac{X + X^T}{2}$$

is a symmetric random matrix, and the probability distribution of such a symmetric matrix is called the **Gaussian Orthogonal Ensemble (GOE)**.

Example 7.7.

$$H = \left(\begin{array}{cc} -0.36091 & 0.475152 \\ 0.475152 & -0.262286 \end{array}\right)$$

is an example of a matrix in a GOE, where $H = \frac{X+X^T}{2}$ is created from $X_{i,j} \sim$
$N(0, 1)$

$$X = \left(\begin{array}{cc} -0.36091 & 0.380648 \\ 0.569655 & -0.262286 \end{array}\right).$$

Lemma 3. *The joint distribution of the elements of an $n \times n$ random matrix $H \sim GOE$ where $H_{i,j} \sim N(0, 1)$ when $i = j$ and $H_{i,j} \sim N(0,\frac{1}{2})$ otherwise is*

$$p(H) = (2\pi)^{-n/2}(\pi)^{-n(n-1)/2}e^{-Tr(H^2)/2}.$$

Proof. The joint distribution of $H_{i,j}$, $1 \le i, j \le n$,

$$\begin{aligned} p(H) &= \prod_{i=1}^{n} \frac{1}{\sqrt{2\pi}} e^{-\frac{H_{i,i}^2}{2}} \cdot \prod_{1 \le i < j \le n} \frac{1}{\sqrt{\pi}} e^{-H_{i,j}^2} \\ &= (2\pi)^{-n/2}(\pi)^{-n(n-1)/2}e^{-\frac{1}{2}\sum_{i=1}^{n} H_{i,i}^2 + 2\sum_{1 \le i < j \le n} H_{i,j}^2} \\ &= (2\pi)^{-n/2}(\pi)^{-n(n-1)/2}e^{-\frac{1}{2}\sum_{i=1}^{n} H_{i,i}^2 + \sum_{i \ne j} H_{i,j}^2} \\ &= (2\pi)^{-n/2}(\pi)^{-n(n-1)/2}e^{-\frac{1}{2}\sum_{i,j=1}^{n} H_{i,j}^2} \end{aligned}$$

Since H is symmetric,

$$\begin{aligned} p(H) &= (2\pi)^{-n/2}(\pi)^{-n(n-1)/2}e^{-\frac{\sum_{i=1}^{n}\sum_{j=1}^{n} H_{j,i}H_{i,j}}{2}} \\ &= (2\pi)^{-n/2}(\pi)^{-n(n-1)/2}e^{-\frac{\sum_{i=1}^{n}(H^2)_{i,i}}{2}} \\ &= (2\pi)^{-n/2}(\pi)^{-n(n-1)/2}e^{-\frac{Tr(H^2)}{2}}. \end{aligned}$$

$\square$

Lemma 4. *Let H be an $n \times n$ GOE matrix and O an $n \times n$ non-random orthogonal matrix. Then the distribution of H is the same as the distribution of OHO^T.*

Proof. First note that since matrix O is an orthogonal matrix,

$$Tr(OHO^T)^2 = Tr(OH^2O^T) = Tr(H^2).$$

The function $f : H \to OHO^T$ has a Jacobian matrix whose determinants are relative to the entries of H. That is, the entries of the Jacobian, denoted $J(O)$, are

$$\partial(OHO^T)_{i,j} \partial H_{k,l} = O_{i,k} O_{j,l}.$$

It is important to note that

$$|J(O)| = |J(O^T)|.$$

Therefore,

$$|J(O)|^2 = |J(O)||J(O^T)| = |J(OO^T)| = |J(I)| = 1.$$

$\square$

The previous two lemmas tell us about the joint distribution of the elements of a GOE matrix. More interestingly, we will further explore the distribution of the eigenvalues and spacing of eigenvalues of random matrices from the GOE. For a general 2×2 random matrix from the GOE,

$$X = \begin{pmatrix} X_1 & X_2 \\ X_2 & X_3 \end{pmatrix},$$

where $X_1, X_3 \sim N(0, 1)$ and $X_2 \sim N(0, \frac{1}{2})$ are i.i.d. The eigenvalues of X are

$$\frac{1}{2} Tr(X) \pm \sqrt{(X_1 - X_3)^2 + 4X_2^2}.$$

Thus the mean of the eigenvalues is $X_1 + X_3$.
The probability density function of $X_1 + X_3 \sim N(0, 2)$, which can be seen in Figure 7.3.

Theorem 16. *(Wigner's Semicircle Law) If X is an $n \times n$ matrix whose entries $X_{ij} \sim N(0, \sigma^2)$,*

$$H = \frac{X + X^T}{\sqrt{2n}},$$

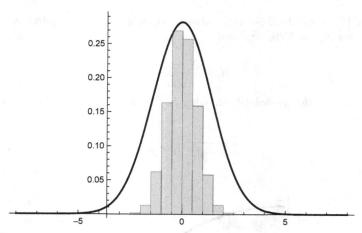

FIGURE 7.3
Distribution of mean of eigenvalues in 2×2 GOE.

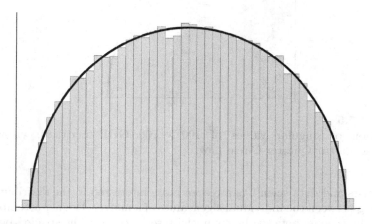

FIGURE 7.4
Distribution of eigenvalues of 100×100 GOE matrices, visualizing Wigner's Semicircle Law.

then as $n \to \infty$, the probability density function for the eigenvalues of H approaches

$$\frac{1}{2\pi}\sqrt{4 - x^2}, \quad -2 \leq x \leq 2.$$

as seen in Figure 7.4.

The **Marchenko-Pastur Quarter-Circle Law** is the analogous property for the singular values of a matrix.

Theorem 17. *(Marchenko-Pastur Quarter-Circle Law) If X is an $n \times n$ matrix whose entries $X_{ij} \sim N(0, \sigma^2)$ and*

$$H = \frac{X + X^T}{\sqrt{2n}},$$

then as $n \to \infty$, the probability density function for the singular values of H approaches

$$\frac{1}{2\pi}\sqrt{4 - x^2}, \ 0 \le x \le 2.$$

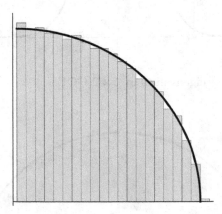

FIGURE 7.5
Distribution of singular values of 100×100 GOE matrices, visualizing Marchenko-Pastur Quarter-Circle Law.

Scientists and data scientists are even more interested in the distribution of the spacing between the individual eigenvalues of a matrix. Wigner attempted to compare the spacing between heavy nuclei resonances and the spacing between eigenvalues of GOE matrices.

This theory, called **Wigner's surmise**, describes a phenomenon called 'level repulsion'. In a spatially bound system, particles can only have discrete amounts of energy, which tend to cluster in a general sense.

In general, Wigner's surmise states that the probability density function of the eigenvalue spacings, S, of a random GOE matrix follows a Rayleigh distribution.

$$p(s) = P(S = s) = \frac{s^2}{k^2}e^{\frac{-s^2}{2k^2}},$$

where k is a scale parameter for the distribution.

Since the entries of such matrices are random variables and not fixed numbers, we must investigate techniques to theoretically determine the probability distribution of the eigenvalues.

We dig deeper into the theoretical aspects of Wigner's surmise for 2×2 GOE random matrices.

If we define a 2×2 random matrix from the GOE,

$$X = \begin{pmatrix} X_1 & X_2 \\ X_2 & X_3 \end{pmatrix},$$

where $X_1, X_3 \sim N(0,1)$ are i.i.d and $X_2 \sim N(0,\frac{1}{2})$, the eigenvalues of X are

$$\frac{1}{2}Tr(X) \pm \sqrt{(X_1 - X_3)^2 + 4X_2^2}.$$

Let $s = \lambda_1 - \lambda_2 = \sqrt{(X_1 - X_3^2) + 4X_2^2}$. The probability density function for s,

$$p(s) = \frac{1}{2\pi^{3/2}} \int_{-\infty}^{\infty} \int_{-\infty}^{\infty} \int_{-\infty}^{\infty} e^{\frac{-(X_1^2 + 2X_2^2 + X_3^2)}{2}} dX_1 dX_2 dX_3$$
$$\delta(d - \sqrt{(X_1 - X_3)^2 + 4X_2^2}).$$

A visualization of $p(s)$ can be seen in Figure 7.6.

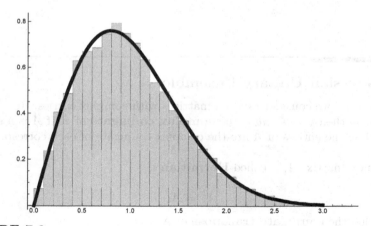

FIGURE 7.6
Example distribution of Wigner Surmise with 2×2 GOE distribution.

With a change of variables,

$$X_1 - X_3 = r\cos(\theta),$$
$$X_1 + X_3 = \eta,$$
$$X_2 = r\sin(\theta)/2,$$

we get,

$$J = \begin{vmatrix} \cos(\theta)/2 & -r\sin(\theta)/2 & 1/2 \\ -\cos(\theta)/2 & r\sin(\theta)/2 & 1/2 \\ \sin(\theta)/2 & \cos(\theta)/2 & 0 \end{vmatrix} = -r/4,$$

$$p(s) = \frac{1}{8\pi^{3/2}} \int_0^\infty \int_0^{2\pi} \int_{-\infty}^\infty e^{\left(-\frac{\left(\frac{r\cos(\theta)+\eta}{2}\right)^2 + \left(\frac{-r\cos(\theta)+\eta}{2}\right)^2 + \frac{r^2\sin^2(\theta)}{2}}{2}\right)} r\delta(d-r)\,d\eta\,d\theta\,dr$$

$$= \frac{1}{8\pi^{3/2}} \int_0^\infty r\,dr\,\delta(s-r) \int_0^{2\pi} d\theta \int_{-\infty}^\infty e^{-\frac{1}{2}\left(r+\frac{\eta^2}{4}\right)}\,d\eta$$

$$= \frac{s^2}{2} e^{-s^2/4}.$$

7.4 Gaussian Unitary Ensemble

In this section, we consider random matrices with complex entries. Recall that if $z = a + ib$ then $\bar{z} = a - ib$ is the **complex conjugate** of z. If A is a square matrix, then the entries of $\bar{A}$ are the complex conjugate of each corresponding entry in A.

A square matrix, A, is called **Hermitian** if

$$\bar{A}^T = A.$$

$\bar{A}^T$ is called the **conjugate transpose** of A.

Example 7.8.

$$A = \begin{pmatrix} 1 & 2+i & 3-2i \\ 2-i & 4 & 7+4i \\ 3+2i & 7-4i & 6 \end{pmatrix}$$

is Hermitian. Notice that the main diagonal entries of a matrix must be real in order for the matrix to be Hermitian. All real symmetric matrices are also Hermitian.

An invertible square matrix, A, is **unitary** if

$$\bar{A}^T = A^{-1}.$$

Example 7.9. *If*

$$B = \frac{1}{2}\begin{pmatrix} 1+i & 1-i \\ 1-i & 1+i \end{pmatrix}$$

$$\bar{B}^T = \frac{1}{2}\begin{pmatrix} 1-i & 1+i \\ 1+i & 1-i \end{pmatrix}.$$

$\bar{B}^T B = I$ *and thus matrix B is unitary; however B is not Hermitian. The matrix*

$$Y = \begin{pmatrix} 0 & \frac{1}{\sqrt{2}} + \frac{i}{\sqrt{2}} \\ \frac{1}{\sqrt{2}} - \frac{i}{\sqrt{2}} & 0 \end{pmatrix}.$$

is an example of a unitary Hermitian matrix.

It is important to note that if an $n \times n$ matrix is an unitary Hermitian matrix then its rows (and columns) form an orthonormal basis for $\mathbb{C}^n$.

The inner product of the columns of Y, under a standard Euclidean inner product, are

$$\left\langle \left(0, \frac{1}{\sqrt{2}} - \frac{i}{\sqrt{2}}\right), \left(\frac{1}{\sqrt{2}} + \frac{i}{\sqrt{2}}, 0\right) \right\rangle = 0,$$

$$\left\| \left(0, \frac{1}{\sqrt{2}} - \frac{i}{\sqrt{2}}\right) \right\| = \left\| \left(\frac{1}{\sqrt{2}} + \frac{i}{\sqrt{2}}, 0\right) \right\| = 1.$$

Every Hermitian matrix, H, is diagonalizable with a unitary matrix, U, and the resulting diagonal matrix, D, will have real entries. That is if H is a Hermitian matrix then there exists a unitary matrix U such that

$$D = U^{-1}HU,$$

where $D_{i,i} \in \mathbb{R}$.

$$H = \begin{pmatrix} 1 & 2+3I \\ 2-3I & 4 \end{pmatrix} \text{ and } U = \begin{pmatrix} (2-3i)\sqrt{\frac{1}{26} - \frac{3}{26\sqrt{61}}} & \frac{\sqrt{61}+3}{\sqrt{6\sqrt{61}+122}} \\ (-2+3i)\sqrt{\frac{1}{26} + \frac{3}{26\sqrt{61}}} & \frac{\sqrt{61}-3}{\sqrt{122-6\sqrt{61}}} \end{pmatrix}$$

where the columns of U are the unit eigenvectors of H, then

$$U^{-1}HU = \begin{pmatrix} \frac{1}{2}\left(\sqrt{61}+5\right) & 0 \\ 0 & \frac{1}{2}\left(5-\sqrt{61}\right) \end{pmatrix} = D.$$

U is a unitary matrix and the main diagonal entries of D are the real eigenvalues of H.

Applying this idea with similar ideas to that in Lemma 4, the probability density of a Hermitian matrix, H, is the same as that for $D = U^{-1}HU$, and thus is directly related to the probability density of the eigenvalues of H.

A random Hermitian matrix is a matrix in the **Gaussian Unitary Ensemble (GUE)** if $H_{i,i} \sim N(0,1)$ and the upper triangular entries $H_{i,j} \sim N\left(0,\frac{1}{2}\right) + iN\left(0,\frac{1}{2}\right)$, for $1 \le i \le j \le n$. The standard GUE,

$$H = \frac{X + \bar{X}^T}{2},$$

where $X_{i,i} \sim N(0,1)$ and $X_{i,j} \sim \mathbf{N}\left(0,\frac{1}{2}\right) + i\mathbf{N}\left(0,\frac{1}{2}\right)$ for $1 \le i \le j \le n$.

Example 7.10. *One 3×3 GUE is*

$$H = \begin{pmatrix} -0.168 & 0.042 + 0.405i & 0.464 - 0.5585i \\ 0.042 - 0.405i & -0.689 & -0.2945 - 0.2925i \\ 0.464 + 0.5585i & -0.2945 + 0.2925i & -0.131 \end{pmatrix}$$

Notice that H is a Hermitian matrix but not unitary. H was created from

$$X = \begin{pmatrix} -0.168 + 1.909i & -0.414 + 0.173i & 0.878 - 0.551i \\ 0.498 - 0.637i & -0.689 + 0.403i & -0.2 - 0.988i \\ 0.05 + 0.566i & -0.389 - 0.403i & -0.131 - 1.189i \end{pmatrix}.$$

Like the Gaussian Orthogonal Ensemble, if we define a 2×2 random matrix from a GUE,

$$X = \begin{pmatrix} X_1 & X_2 + X_3 i \\ X_2 - X_3 i & X_4 \end{pmatrix},$$

with i.i.d random variables X_1, $X_4 \sim N(0,1)$ and X_2, $X_3 \sim N\left(0,\frac{1}{2}\right)$, the eigenvalues of X are

$$\frac{1}{2}Tr(X) \pm \sqrt{(X_1 - X_4)^2 + 4(X_2 + X_3 i)^2},$$

and the mean of the eigenvalues is $\frac{1}{2}\left(X_1 + X_4\right) \sim N\left(0,\frac{1}{2}\right)$.

Theorem 18. *The eigenvalues of a Hermitian matrix are all real and eigenvectors of distinct eigenvalues are orthogonal.*

Proof. Let A be a Hermitian matrix with eigenvalue λ and corresponding eigenvector x, then

$$Ax = \lambda x.$$

Let $\alpha = x^T A x$. Then α is a real number since

$$\overline{\alpha} = \alpha^T = (x^T A x)^T = x^T A^T x = x^T A x = \alpha.$$

Additionally, $\alpha = \lambda x^T x$, so λ is also a real number.

If x_1 and x_2 are eigenvectors corresponding to distinct eigenvalues λ_1 and λ_2,

$$(Ax_1)^T x_2 = x_1^T (A^T x_2) = x_1^T (A x_2) = x_1^T (\lambda_2 x_2) = \lambda_2 x_1^T x_2.$$

Similarly,

$$(Ax_1)^T x_2 = \lambda_1 x_1^T x_2.$$

Since $\lambda_1 \neq \lambda_1$ then $x_1^T x_2 = 0$ and thus x_1 and x_2 are orthogonal. $\qquad\square$

The result presented in Theorem 18 is not necessarily true for all random matrices which are not GOE or GUE. Figure 7.7 shows the real versus imaginary part of the mean eigenvalue of a non-Hermitian, non-symmetric, random matrix.

The final ensemble of this chapter is the Gaussian Symplectic Ensemble, which focuses on random entries who are quaternion in nature.

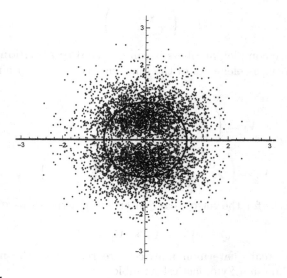

FIGURE 7.7
The Real vs Imaginary parts of the mean eigenvalue of a 2×2 Non-Hermitian Random Matrix.

7.5   Gaussian Symplectic Ensemble

A **quaternion** is a linear combination

$$q = a \cdot 1 + bi + cj + dk,$$

where a, b, c, and $d \in \mathbb{R}$ and i, j, and k are the quaternion units,

$$i^2 = j^2 = k^2 = i \cdot j \cdot k = -1.$$

The set of quaternions is denoted $\mathbb{H}$.

The complex conjugate for a quaternion, q, can also be defined

$$\bar{q} = a \cdot 1 - bi - cj - dk.$$

Similar to elements in $\mathbb{C}^n$, if $q = (q_1, q_2, \ldots, q_n)$ and $p = (p_1, p_2, \ldots, p_n)$ are in $\mathbb{H}^n$ then

$$< p, q >= \sum_{i=1}^{n} \bar{p}_i q_i,$$

and thus the norm of q,

$$||q|| = q\bar{q} = \bar{q}q = a^2 + b^2 + c^2 + d^2.$$

A 2×2 matrix, Q, of the form

$$Q = \begin{pmatrix} z & w \\ -\bar{w} & \bar{z} \end{pmatrix},$$

where z and w are complex numbers is called a **real quaternion** matrix. Just like the standard basis elements of the quaternions, are 1, i, j and k, as 2×2 matrices

$$V_1 = \begin{pmatrix} 1 & 0 \\ 0 & 1 \end{pmatrix}, V_2 = \begin{pmatrix} i & 0 \\ 0 & -i \end{pmatrix},$$

$$V_3 = \begin{pmatrix} 0 & 1 \\ -1 & 0 \end{pmatrix}, \text{ and } V_4 = \begin{pmatrix} 0 & i \\ i & 0 \end{pmatrix},$$

can serve as a basis for the set of real quaternion matrices where

$$Q = aV_1 + bV_2 + cV_3 + dV_4.$$

We will see how real quaternion matrices are related to the next ensemble presented, the Gaussian Symplectic Ensemble.

A matrix A is **skew-symmetric** if

$$A^T = -A$$

Example 7.11.

$$A = \begin{pmatrix} 0 & 2 & 3 \\ -2 & 0 & 4 \\ -3 & -4 & 0 \end{pmatrix} \text{ and } B = \begin{pmatrix} 0 & -i & -2i \\ i & 0 & 3i \\ 2i & -3i & 0 \end{pmatrix}.$$

Matrix A is an example of skew-symmetric matrix and matrix B is an example of a skew-symmetric Hermitian matrix.

A **symplectic** matrix M is a $2n \times 2n$ matrix such that if $S = \begin{pmatrix} 0 & I_n \\ -I_n & 0 \end{pmatrix}$ is a real skew-symmetric matrix then

$$M^T S M = S.$$

Example 7.12.

$$A = \begin{pmatrix} 0 & 0 & 1 & 0 \\ 0 & 0 & 0 & 1 \\ -1 & 0 & 0 & 0 \\ 0 & -1 & 0 & 0 \end{pmatrix} \text{ and } B = \begin{pmatrix} 0 & i & 0 & 0 \\ -i & 0 & 2i & 0 \\ 0 & -2i & 0 & -i \\ 0 & 0 & i & 0 \end{pmatrix}$$

Matrix A is and example of a symplectic matrix, where matrix B is a symplectic, skew-symmetric, and Hermitian matrix.

If X is a $2n \times 2n$ matrix with entries $X_{i,j} \sim N(0, \sigma^2) + iN(0, \sigma^2)$ and $X_{i,j}$ are i.i.d. then we define

$$H = \frac{X + \bar{X}^T - S(X + \bar{X}^T)^T S}{4}$$

as a **Gaussian Symplectic Ensemble (GSE)**, where $S = \begin{pmatrix} 0 & -1 \\ 1 & 0 \end{pmatrix} \otimes I_n$.

Example 7.13. *If*

$$X = \begin{pmatrix} 0 & i & 0 & i \\ 0 & 2 & 0 & -1+i \\ -i & -2i & -1 & -1+i \\ -1-i & 1 & 1 & 2i \end{pmatrix}$$

$$\text{and } X + \bar{X}^T = \begin{pmatrix} 0 & i & i & -1+2i \\ -i & 4 & 2i & i \\ -i & -2i & -2 & i \\ -1-2i & -i & -i & 0 \end{pmatrix}.$$

$$S = \begin{pmatrix} 0 & -1 \\ 1 & 0 \end{pmatrix} \otimes \begin{pmatrix} 1 & 0 \\ 0 & 1 \end{pmatrix}$$

$$= \begin{pmatrix} 0 & 0 & -1 & 0 \\ 0 & 0 & 0 & -1 \\ 1 & 0 & 0 & 0 \\ 0 & 1 & 0 & 0 \end{pmatrix}$$

Then the matrix

$$H = \frac{X + \bar{X}^T - S(X + \bar{X}^T)^T S}{4}$$

$$= \begin{pmatrix} \frac{1}{2} & 0 & 0 & -\frac{1}{4} \\ 0 & 1 & -\frac{1}{4} & 0 \\ 0 & -\frac{1}{4} & -\frac{1}{2} & 0 \\ -\frac{1}{4} & 0 & 0 & -1 \end{pmatrix}.$$

We have seen three different types of matrix ensembles. You can find a deep dive into the joint distributions of elements of the GOE and that of the eigenvalues of the GOE in Section 7.3. In general if $\beta = 1$, 2, and 4, represent the GOE, GUE, and GSE, the joint distribution of the elements from a random matrix, X, are of the form

$$\frac{1}{2^{n/2}} \frac{1}{\pi^{n/2+n(n-1)\beta/4}} e^{-||X||_F^2}$$

where $|| \cdot ||_F$ represents the **Frobenius norm**,

$$||X||_F^2 = \sum_{i=1}^{n} \sum_{j=1}^{n} |x_{i,j}|^2.$$

We now look at the importance of random matrices to a variety of applications.

7.6 Random Matrices and the Relationship to the Covariance

There are some particular random matrices of interest that we will explore in this section. We first define a real symmetric $m \times m$ random matrix

$$M = XX^T = \sum_{j=1}^{n} X_j X_j^T$$

where X is a $m \times n$ random matrix, with columns X_j, $1 \le j \le n$, are formed from independent samples of size m from i.i.d. random variables. The matrix M is called a **real Wishart matrix**. If these random variables are Gaussian each with mean 0 and covariance

$$\Sigma = \sum (X - \bar{X})(X - \bar{X})^T,$$

then M is a **Gaussian Wishart matrix** and is denoted $W_p(n,\Sigma)$, and has a probability distribution,

$$f(M) = \frac{1}{(2\pi)^{mn/2}\Gamma_p(\frac{n}{2})|\Sigma|^{n/2}}|M|^{(n-p-1)/2}e^{-\frac{1}{2}Tr(\Sigma^{-1}M)}.$$

Here $\Gamma_p(n)$ is the multivariate gamma function.

$$\Gamma(x) = \int_0^\infty y^{x-1}e^{-y}dy.$$

Notice that since M is symmetric, all of the eigenvalues, $\lambda_1, \lambda_2, \ldots, \lambda_m$ of M are real.

In Section 3.5, we talked about the covariance and correlation matrices. Given a data set, X, the empirical covariance matrix can be found by finding

$$XX^T.$$

The empirical covariance matrix is different than the true, statistical, correlation matrix, C. If there are a large number of observation then, in fact, the empirical covariance matrix will be to the true covariance matrix.

Example 7.14. *Let $Z = (X, Y)^T$ where $X \sim N(0, \sigma_x^2)$, $Y \sim N(0, \sigma_y^2)$. Then the covariance matrix* $\Sigma = \begin{pmatrix} \sigma_x^2 & \rho\sigma_x\sigma_y \\ \rho\sigma_x\sigma_y & \sigma_y^2 \end{pmatrix}$ *and*

$$M = ZZ^T = \begin{pmatrix} \sum_i X_i^2 & \sum_i X_i Y_i \\ \sum_i X_i Y_i & \sum_i Y_i^2 \end{pmatrix}$$

where ρ is the correlation coefficient.

7.7   CASE STUDY: Finance and Brownian Motion

In this case study, we will discuss how the covariance and correlation matrices relate to random matrices through a financial lens.

The unconditional probability distribution for any traded financial commodity (such as stocks, currencies, interest rates) is far from a Gaussian distribution. This unconditional probability distribution tends to follow a distribution similar to a **Rayleigh distribution** or more generally a **power law** distribution, called the **Tracy-Widom distribution**.

The general form of a Rayleigh probability density function is

$$f(x) = \frac{x}{\sigma^2} e^{\frac{-x^2}{2\sigma^2}}, \, 0 \leq x < \infty,$$

where $E(X) = \sigma\sqrt{\frac{\pi}{2}}$ and $Var(X) = \left(\frac{4-\pi}{2}\right)\sigma^2$.
Some examples of this probability distribution can be see in Figure 7.8.

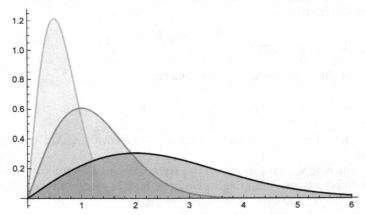

FIGURE 7.8
Examples of Rayleigh Distributions with different scaling factors.

The largest eigenvalue of a random Gaussian Ensemble matrix and its corresponding eigenvector can be interpreted as the collective response of the market to external factors, so it can be compared with the market index.

If H is a GUE random matrix, Figure 7.9 shows an example of what the distribution of maximum eigenvalues of H would look like.

Stocks have quite a bit of randomness in their volatility which can been seen in the three airline stock trends seen in Figure 7.10. Table 7.1 shows a small subset of the data for these three airline stocks, where columns represent the

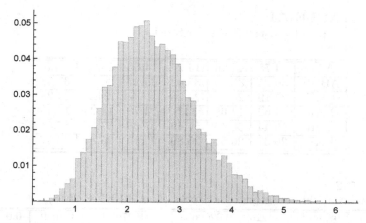

FIGURE 7.9
Distribution of maximum eigenvalues of a GUE matrix.

trading days and rows for the different stocks. The stocks in this problem are correlated.

We begin by manipulating the data in Table 7.1, as we have done in other data sets throughout the text. Here we apply a logarithmic transformation of the daily returns, which can be seen in Table 7.2.

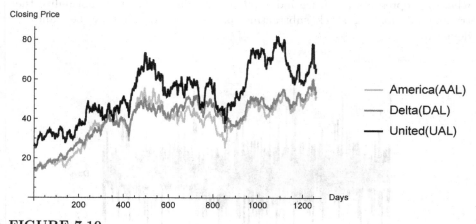

FIGURE 7.10
American, Delta, and United Airline stock closing prices February 2013–February 2018.

TABLE 7.1

Sample of Closing Prices for Stocks (Feb 2013–Feb 2018).

AAL	14.75	14.46	14.27	14.66	13.99	14.5
ABBV	36.25	35.85	35.42	35.27	36.57	37.58
CBG	24.19	24.13	24.26	24.5	24.7	24.28
DAL	14.62	14.69	14.5	14.78	14.24	14.45
EQR	55.44	56.46	57.49	57.65	57.56	57.51
UAL	26.31	26.51	25.89	26.19	25.87	26.37

TABLE 7.2

Sample of Logarithmic Returns for Stocks (Feb 2013–Feb 2018).

AAL	-0.0198569	-0.0132268	0.0269633	-0.0467799	0.0358059	-0.0166902
ABBV	-0.0110958	-0.0120669	-0.00424389	0.0361955	0.0272438	0.0161017
CBG	-0.00248344	0.00537302	0.00984421	0.00813013	-0.0171503	0.00902015
DAL	0.00477654	-0.0130183	0.0191263	-0.03722	0.0146395	-0.0104349
EQR	0.0182311	0.0180786	0.00277923	-0.00156236	-0.000869036	0.00121644
UAL	0.00757293	-0.0236652	0.0115209	-0.0122937	0.019143	0.0143075

First, we will calculate the **logarithmic returns** for consecutive trading days for each stock.

$$\log(return) = \log\left(\frac{p_i}{p_{i-1}}\right),$$

where p_i represents the price index on trading day i. Then we normalize the trading data for each stock, subtracting the mean and dividing by the standard deviation.

FIGURE 7.11

Logarithmic returns for American Airline stock February 2013–February 2018.

In this case study, we will limit our study to small correlated data sets representing a 6 day period. A 6×6, *stock* $\times$ *log(return)*, matrix, X, can be defined from the data in Table 7.2. Notice that this matrix uses the logarithmic returns of the 6 stocks over the first six day period, Day 1 through Day 6. A similar matrix can be created from Day 2 through Day 7 and in general in the stock price observations are taken for M days then the matrix will have M columns representing Day i through Day $i + M$.

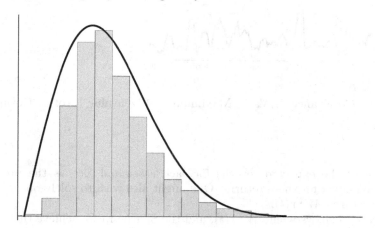

FIGURE 7.12
Distribution of Largest Eigenvalues for the Correlation Matrix.

For each M consecutive days, we can create the correlation matrix

$$W = \frac{XX^T}{M},$$

and determine the eigenvalues of the correlation matrix W. Notice that W is a Wishart matrix.

Figure 7.12 shows the distribution of the largest eigenvalue of W using a larger dataset from the stocks in Table 7.2 from February 2013 to February 2018, which can be found at Github link 48.

A covariance matrix relating the log returns over a 6 day period, can also help us understand the behaviors of stocks in Table 7.11.

We can also look at how the largest eigenvalues of W throughout time are related to the average covariance as well. Notice the behavior similarities in Figure 7.13.

In order to create a optimal portfolio, one might want to maximize the **Sharpe ratio** [38],

$$\frac{\text{Portfolio return}}{\text{Portfolio risk}} = \frac{W^T r(t)}{\sqrt{W^T \Sigma(t) W}},$$

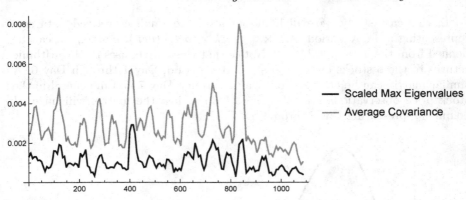

FIGURE 7.13
Average   Covariance   Vs   Maximum   Eigenvalue   of   Covariance
Matrix.

where $r(t)$ is the expected returns for each asset and $\Sigma(t)$ is the predicted covariance matrix for these returns. One might also wish to solely optimize the expected returns, $W^T r(t)$.

Figure 7.14 shows where $W^T r(t)$, and the Sharpe ratio, achieve their maximum throughout time. Notice that maximums occur roughly every 20 days.

Github links 44 and 46 provide Python and R syntax for this case study.

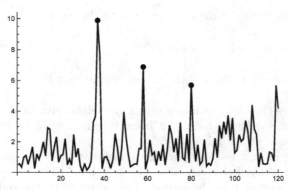

FIGURE 7.14
$W^T r(t)$ throughout time.

7.8 CASE STUDY: Random Matrices in Gene Interaction

In reality, genes do not act in isolation within the body but interact within a common cellular environment. This could be a dominant gene interaction, a complementary gene interaction, or a supplementary gene interaction. Scientists have known for some time that some genetic diseases result from mutations of a single gene. These diseases are called **monogenetic diseases**. **Multigenetic diseases** results from genetic mutations of more than one gene that interact with one another. This process is called **epistatis**. Knowledge about gene-gene interactions can give us a better understanding about hereditary diseases [29].

There are many databases related to gene-gene interaction. For this case study, we use data from The Cancer Network Galaxy [5], which focuses on gene-gene interaction related to cancer. The Cancer Network Galaxy has 256 data sets with 22,820 genes and their interactions represented. In this case study, we focus on one such data set focusing on mRNA expression of 131 cancer cell lines that can be found at Github link 49; however, a larger model incorporating multiple data sets is encouraged for future study.

Figure 7.15 shows a graphical depiction of a subgraph of interactions between genes in the dataset.

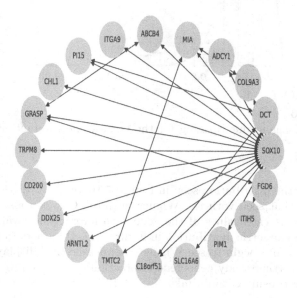

FIGURE 7.15
Visualization of a small subset of gene-gene interactions in the Cancer Network Galaxy.

For this case study, assume that there are n genetic makers, $G_1, G_2, \ldots, G_n$ measured on m subjects. In the provided data set there is an identified parent gene and child gene. A relationship matrix M is created such that if parent gene G_i is related to child gene G_j then $M_{i,j} = 1$.

Otherwise $M_{i,j} = 0$. Note that the matrix M is a sparse matrix with many entries equal to 0. Also notice that the matrix M may not be symmetric.

We define the sample correlation matrix

$$W = \frac{1}{n-1} M M^T.$$

In addition, if λ_i and λ_j are consecutive eigenvalues of W, Figure 7.16 shows the distribution of the spacing between consecutive eigenvalues.

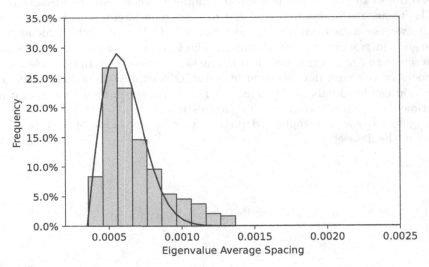

FIGURE 7.16
Distribution of Eigenvalue Spacing for the Gene-Gene Interaction Correlation Matrix.

Notice that the distribution seen in Figure 7.16 follows a Rayleigh distribution similar to that seen with the Wigner Surmise. One can also see that the distribution of the eigenvalues of the gene-gene interaction covariance matrix follows a more symmetric distribution in Figure 7.17. If more data from [5] were integrated, we would begin to see Wigner's Semicircle Law displayed.

In order to further study gene-gene interaction relationship to random matrices, you may consult Github links 45 and 47 for Python and R coding suggestions.

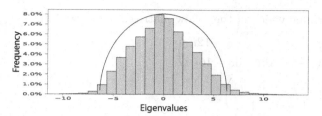

FIGURE 7.17
Distribution of Eigenvalues of Gene-Gene Interaction Correlation Matrix.

7.9 Exercises

1. Find the condition number of $A = \begin{pmatrix} -0.1 & -1 \\ 0 & 1 \end{pmatrix}$.

2. Verify that the Bauer-Fike Theorem holds for

$$A = \begin{pmatrix} 4+i & 0 \\ 3+i & 1 \end{pmatrix} \text{ and } E = \begin{pmatrix} -0.1 & 0.8 \\ -0.01 & 0 \end{pmatrix}.$$

   Recall that $|a + bi| = \sqrt{a^2 + b^2}$.

3. (Coding) The following commands can be used to visualize the pseudospectrum of a matrix A using Python.

   ```
   import numpy as np
   import matplotlib.pyplot as plt
   from scipy.linalg import svdvals ,schur
   ```

   Graph the pseudospectrum of

   $$A = \begin{pmatrix} -0.1 & -1 \\ 0 & 1 \end{pmatrix}.$$

4. a. If H is a 4×4 GOE find the probability distribution of the mean eigenvalues.

    b. (Coding) Write code to generate 10000 4×4 GOE matrices and calculate the mean of the eigenvalues of each GOE. Use these values to plot the probability distribution of the mean eigenvalues and compare it to your answer in part a.

5. Determine which of the follow matrices are Hermitian.

   a. $\begin{pmatrix} 4 - 2i & -8 + 12i \\ -8 + 12i & 36 - 16i \end{pmatrix}$

   b. $\begin{pmatrix} 1 & 1 + i \\ 1 - i & 1 \end{pmatrix}$

   c. $\begin{pmatrix} 1 & 2 \\ 2 & 3 \end{pmatrix}$

6. Let $A = \begin{pmatrix} \frac{1+i}{2} & -\frac{1+i}{2} \\ \frac{1}{\sqrt{2}} & \frac{1}{\sqrt{2}} \end{pmatrix}$. Determine if A is a unitary matrix.

7. (Coding) Using either Python or R, generate a population of $N = 100$ people that have all the same initial value $V = 1000$. Write a loop, where the time step represents trading periods, and keep track of each person's value under the following conditions.

   a. Random trading with one person at each time step, total value between the two traders is shared equally between the traders.

   b. Random trading with one person at each time step, total value between the two traders is shared where the percent split is also generated randomly.

   c. Each person is given a random initial value between $V = 100$ and $V = 1000$. Random trading with one person at each step. The person of greater value gets 75% of the total value between the two traders and the lesser get 25%.

   d. Create a probability histogram representing the values of the population members in parts a. through c. and estimate the distribution for each part.

   e. For part c., create a matrix, M, where the rows represent the population members and the columns represent the individual values throughout time. Use your matrix M to create a correlation matrix.

   f. Create a probability histogram representing the eigenvalues of the correlation matrix in part e. and estimate the distribution.

8. (Coding) Create code in Python or R to generate the average condition number of a 2×2 GOE matrix, A, and the average error in a system $Ax = b$.

8

Sample Solutions to Exercises

8.1 Chapter 1

1. a. $\sum_{j=2}^{6} A_{1,j} = 2$
    b. $A_{1,j}^2 \neq 0$ for $i = 2,3,4,5,6$ so Kush has connections with all of the other 5 in 2 steps.
    c. $A_{1,2} + A_{1,2}^2 = 2$

2. $\rho(A) = 1 + \sqrt{2}$.

3. The graph is not a simple connected graph since it contains a loop.

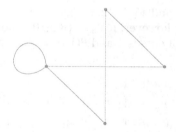

4. a. $A = \begin{pmatrix} 0 & 1 & 0 & 0 & 0 & 0 & 1 & 1 & 0 & 0 \\ 1 & 0 & 1 & 0 & 0 & 0 & 0 & 0 & 0 & 0 \\ 0 & 1 & 0 & 1 & 0 & 0 & 0 & 0 & 0 & 0 \\ 0 & 0 & 1 & 0 & 1 & 0 & 0 & 0 & 0 & 0 \\ 0 & 0 & 0 & 1 & 0 & 1 & 0 & 0 & 0 & 0 \\ 0 & 0 & 0 & 0 & 1 & 0 & 1 & 0 & 0 & 0 \\ 1 & 0 & 0 & 0 & 0 & 1 & 0 & 0 & 0 & 0 \\ 1 & 0 & 0 & 0 & 0 & 0 & 0 & 0 & 1 & 1 \\ 0 & 0 & 0 & 0 & 0 & 0 & 0 & 1 & 0 & 1 \\ 0 & 0 & 0 & 0 & 0 & 0 & 0 & 1 & 1 & 0 \end{pmatrix}$

DOI: 10.1201/9781003025672-8

b. $L =$

$$\begin{pmatrix} 3 & -1 & 0 & 0 & 0 & 0 & -1 & -1 & 0 & 0 \\ -1 & 2 & -1 & 0 & 0 & 0 & 0 & 0 & 0 & 0 \\ 0 & -1 & 2 & -1 & 0 & 0 & 0 & 0 & 0 & 0 \\ 0 & 0 & -1 & 2 & -1 & 0 & 0 & 0 & 0 & 0 \\ 0 & 0 & 0 & -1 & 2 & -1 & 0 & 0 & 0 & 0 \\ 0 & 0 & 0 & 0 & -1 & 2 & -1 & 0 & 0 & 0 \\ -1 & 0 & 0 & 0 & 0 & -1 & 2 & 0 & 0 & 0 \\ -1 & 0 & 0 & 0 & 0 & 0 & 0 & 3 & -1 & -1 \\ 0 & 0 & 0 & 0 & 0 & 0 & 0 & -1 & 2 & -1 \\ 0 & 0 & 0 & 0 & 0 & 0 & 0 & -1 & -1 & 2 \end{pmatrix}$$

c. The Fiedler value is 0.0.237487 and the unit Fiedler vector is

$$\{0.05213, -0.1147, -0.2543, -0.3335, -0.3335, -0.2543,$$
$$- 0.1147, 0.3734, 0.4898, 0.4898\}.$$

d. Using the Fiedler vector from part c, we can cluster the vertices into two subgroups $V_1 = \{2,3,4,5,6,7\}$ and $V_2 = \{1,8,9,10\}$.

5. If $V_2 = \{1,8,9,10\}$ and $V_1 = \{2,3,4,5,6,7\}$ the number of cut edges is 2 and $\theta(V_2) = \frac{2}{4 \cdot 6}$; however, if $V_2 = \{8,9,10\}$ and $V_1 = \{1,2,3,4,5,6,7\}$ then the number of cuts is 1 and $\theta(V_2) = \frac{1}{3 \cdot 7}$. Thus the isoperimetric ratio is $\theta(G) = \frac{1}{21}$.

6. a. $A =$

$$\begin{pmatrix} 0 & 1 & 1 & 0 & 1 & 1 & 0 & 0 & 0 & 0 & 0 & 0 & 0 & 0 & 0 \\ 1 & 0 & 1 & 0 & 1 & 0 & 0 & 0 & 0 & 0 & 0 & 0 & 0 & 0 & 0 \\ 1 & 1 & 0 & 0 & 1 & 1 & 1 & 0 & 0 & 0 & 0 & 0 & 0 & 0 & 0 \\ 0 & 0 & 0 & 0 & 0 & 1 & 0 & 0 & 0 & 0 & 0 & 0 & 0 & 0 & 0 \\ 1 & 1 & 1 & 0 & 0 & 0 & 1 & 1 & 0 & 0 & 0 & 0 & 0 & 0 & 0 \\ 1 & 0 & 1 & 1 & 0 & 0 & 0 & 1 & 0 & 0 & 0 & 0 & 0 & 0 & 0 \\ 0 & 0 & 1 & 0 & 1 & 0 & 0 & 0 & 1 & 1 & 1 & 0 & 1 & 1 & 0 \\ 0 & 0 & 0 & 0 & 1 & 1 & 0 & 0 & 0 & 0 & 0 & 0 & 0 & 0 & 0 \\ 0 & 0 & 0 & 0 & 0 & 0 & 1 & 0 & 0 & 0 & 1 & 0 & 1 & 0 & 0 \\ 0 & 0 & 0 & 0 & 0 & 0 & 1 & 0 & 0 & 0 & 0 & 0 & 1 & 1 & 0 \\ 0 & 0 & 0 & 0 & 0 & 0 & 1 & 0 & 1 & 0 & 0 & 0 & 1 & 0 & 1 \\ 0 & 0 & 0 & 0 & 0 & 0 & 0 & 0 & 0 & 0 & 0 & 0 & 1 & 1 & 1 \\ 0 & 0 & 0 & 0 & 0 & 0 & 1 & 0 & 1 & 1 & 1 & 1 & 0 & 1 & 0 \\ 0 & 0 & 0 & 0 & 0 & 0 & 1 & 0 & 0 & 1 & 0 & 1 & 1 & 0 & 1 \\ 0 & 0 & 0 & 0 & 0 & 0 & 0 & 0 & 0 & 0 & 1 & 1 & 0 & 1 & 0 \end{pmatrix}.$$

b. Calculating $I + A + A^2$ we can see that no marmot is connected to every other marmot directly or through a two step relationship, however, this calculation shows that marmot 7 is related to every other marmot except for marmot 4 either directly or through a two step relationship. Adding three step relationships allows for marmot 7 to be related to every other marmot in the madness.

c. $L =$

$$\begin{pmatrix}
4 & -1 & -1 & 0 & -1 & -1 & 0 & 0 & 0 & 0 & 0 & 0 & 0 & 0 & 0 \\
-1 & 3 & -1 & 0 & -1 & 0 & 0 & 0 & 0 & 0 & 0 & 0 & 0 & 0 & 0 \\
-1 & -1 & 5 & 0 & -1 & -1 & -1 & 0 & 0 & 0 & 0 & 0 & 0 & 0 & 0 \\
0 & 0 & 0 & 1 & 0 & -1 & 0 & 0 & 0 & 0 & 0 & 0 & 0 & 0 & 0 \\
-1 & -1 & -1 & 0 & 5 & 0 & -1 & -1 & 0 & 0 & 0 & 0 & 0 & 0 & 0 \\
-1 & 0 & -1 & -1 & 0 & 4 & 0 & -1 & 0 & 0 & 0 & 0 & 0 & 0 & 0 \\
0 & 0 & -1 & 0 & -1 & 0 & 7 & 0 & -1 & -1 & -1 & 0 & -1 & -1 & 0 \\
0 & 0 & 0 & 0 & -1 & -1 & 0 & 2 & 0 & 0 & 0 & 0 & 0 & 0 & 0 \\
0 & 0 & 0 & 0 & 0 & 0 & -1 & 0 & 3 & 0 & -1 & 0 & -1 & 0 & 0 \\
0 & 0 & 0 & 0 & 0 & 0 & -1 & 0 & 0 & 3 & 0 & 0 & -1 & -1 & 0 \\
0 & 0 & 0 & 0 & 0 & 0 & -1 & 0 & -1 & 0 & 4 & 0 & -1 & 0 & -1 \\
0 & 0 & 0 & 0 & 0 & 0 & 0 & 0 & 0 & 0 & 0 & 3 & -1 & -1 & -1 \\
0 & 0 & 0 & 0 & 0 & 0 & -1 & 0 & -1 & -1 & -1 & -1 & 6 & -1 & 0 \\
0 & 0 & 0 & 0 & 0 & 0 & -1 & 0 & 0 & -1 & 0 & -1 & -1 & 5 & -1 \\
0 & 0 & 0 & 0 & 0 & 0 & 0 & 0 & 0 & 0 & -1 & -1 & 0 & -1 & 3 \\
\end{pmatrix}.$$

d. If $V_1 = \{v_1, v_2, v_3, v_5, v_6, v_7\}$ then the number of graph cuts is 8 and the isometric ratio is $\frac{4}{27}$.

e. A unit Fiedler vector is approximately
$\{-0.237, -0.216, -0.173, -0.459, -0.169, -0.314, 0.122, -0.287,$
$0.22, 0.225, 0.234, 0.286, 0.235, 0.246, 0.285\}$.

f. Using the result from part d., a cluster of marmots would be
$V_1 = \{v_1, v_2, v_3, v_4, v_5, v_6, v_8\}$ and $V_2 = \{v_7, v_9, v_{10}, v_{11}, v_{12}, v_{13}, v_{14}, v_{15}\}$.

g. Using V_1 and V_2 from part e., we further divide these sets of vertices into subsets using similar analysis. The adjacency matrices associated with V_1 and V_2 respectively are

$$A_1 = \begin{pmatrix}
0 & 1 & 1 & 0 & 1 & 1 & 0 \\
1 & 0 & 1 & 0 & 1 & 0 & 0 \\
1 & 1 & 0 & 0 & 1 & 1 & 0 \\
0 & 0 & 0 & 0 & 0 & 1 & 0 \\
1 & 1 & 1 & 0 & 0 & 0 & 1 \\
1 & 0 & 1 & 1 & 0 & 0 & 1 \\
0 & 0 & 0 & 0 & 1 & 1 & 0 \\
\end{pmatrix}$$

$$\text{and } A_2 = \begin{pmatrix} 0 & 1 & 1 & 1 & 0 & 1 & 1 & 0 \\ 1 & 0 & 0 & 1 & 0 & 1 & 0 & 0 \\ 1 & 0 & 0 & 0 & 0 & 1 & 1 & 0 \\ 1 & 1 & 0 & 0 & 0 & 1 & 0 & 1 \\ 0 & 0 & 0 & 0 & 0 & 1 & 1 & 1 \\ 1 & 1 & 1 & 1 & 1 & 0 & 1 & 0 \\ 1 & 0 & 1 & 0 & 1 & 1 & 0 & 1 \\ 0 & 0 & 0 & 1 & 1 & 0 & 1 & 0 \end{pmatrix}$$

with row/column 7 of A_1 representing marmot 8 and row/column 1 of A_2 representing marmot 7. Finding the Fiedler vector for each of the Laplacian matrices associated with A_1 and A_2 yield a clustering of

$$\{v_1,v_2,v_3,v_5,v_8\},\{v_4,v_5\},\{v_7,v_9,v_{10},v_{11},v_{13}\},\{v_{12},v_{14},v_{15}\}.$$

7. With more knowledge about the marmot interactions, the graph may be a directed graph related to the direction of dominance or could be a weighted graph where the weights are representative of the number of interactions or dominant behaviors that took place.

8.   a. $L =$

$$\begin{pmatrix} 29 & -10 & -3 & 0 & -6 & -10 & 0 & 0 & 0 & 0 & 0 & 0 & 0 & 0 & 0 \\ -10 & 20 & -6 & 0 & -4 & 0 & 0 & 0 & 0 & 0 & 0 & 0 & 0 & 0 & 0 \\ -3 & -6 & 40 & 0 & -3 & -15 & -13 & 0 & 0 & 0 & 0 & 0 & 0 & 0 & 0 \\ 0 & 0 & 0 & 13 & 0 & -13 & 0 & 0 & 0 & 0 & 0 & 0 & 0 & 0 & 0 \\ -6 & -4 & -3 & 0 & 21 & 0 & -1 & -7 & 0 & 0 & 0 & 0 & 0 & 0 & 0 \\ -10 & 0 & -15 & -13 & 0 & 51 & 0 & -13 & 0 & 0 & 0 & 0 & 0 & 0 & 0 \\ 0 & 0 & -13 & 0 & -1 & 0 & 43 & 0 & -5 & -5 & -5 & 0 & -9 & -5 & 0 \\ 0 & 0 & 0 & 0 & -7 & -13 & 0 & 20 & 0 & 0 & 0 & 0 & 0 & 0 & 0 \\ 0 & 0 & 0 & 0 & 0 & 0 & -5 & 0 & 24 & 0 & -11 & 0 & -8 & 0 & 0 \\ 0 & 0 & 0 & 0 & 0 & 0 & -5 & 0 & 0 & 14 & 0 & 0 & -6 & -3 & 0 \\ 0 & 0 & 0 & 0 & 0 & 0 & -5 & 0 & -11 & 0 & 37 & 0 & -11 & 0 & -10 \\ 0 & 0 & 0 & 0 & 0 & 0 & 0 & 0 & 0 & 0 & 0 & 42 & -15 & -12 & -15 \\ 0 & 0 & 0 & 0 & 0 & 0 & -9 & 0 & -8 & -6 & -11 & -15 & 59 & -10 & 0 \\ 0 & 0 & 0 & 0 & 0 & 0 & -5 & 0 & 0 & -3 & 0 & -12 & -10 & 44 & -14 \\ 0 & 0 & 0 & 0 & 0 & 0 & 0 & 0 & 0 & 0 & -10 & -15 & 0 & -14 & 39 \end{pmatrix}.$$

  b. The Fiedler value is 2.38721 and a unit Fiedler vector is approximately $\{-0.27, -0.26, -0.2, -0.35, -0.18, -0.29,$ $-0.02, -0.29, 0.28, 0.28, 0.27, 0.27, 0.23, 0.25, 0.28\}$.

  c. Based on the Fiedler vector, two clusters could be $V_1 = \{v_1,v_2,v_3,v_4,v_5,v_6,v_7,v_8\}$ and $V_2 = \{v_9,v_{10},v_{11},v_{12},v_{13},v_{14},v_{15}\}$.

9.   a. $S = \{-0.333, -1., 1., -0.19, 0.231, -0.487\}$.

  b. With ranking vector $\{-0.626, -2.678, 1.553, -0.484, 0.451, -0.744\}$, marmot 12 is most dominant, with marmot 14 being 2nd and marmot 10, 3rd.

  c. With ranking vector $\{-0.993, -4.905, 1.777, -1.974, 0.381, -1.905\}$, marmot 12 is most dominant, with marmot 14 being 2nd and marmot 10, 3rd.

10. a. If A is the binary *story* × *attribute* matrix. The gramian matrix $S = AA^T$.

b. A (1,1) entry of 3 shows that story 1 has 3 attributes present. A (1,2) entry of 1 says that stories 1 and 2 share 1 attribute in common.

c. $V_1 = \{1, 3, 4, 5, 6, 7, 8, 15, 16, 17, 18, 19, 20, 21, 22, 23, 24, 26, 31\}$ and $V_2 = \{2, 9, 10, 11, 12, 13, 14, 25, 27, 28, 29, 30\}$

11. For $n = 4$ the ranking is (FL, GA, TN, SC, KY, MIZ, VAN) and for $n = 5$ the ranking is (FL, TN, SC, KY, GA, VAN, MIZ)

8.2 Chapter 2

1. A_1 is primitive and A_2 is neither primitive nor irreducible.

2. With transition matrix

$$T = \begin{pmatrix} .8 & 0 & .25 \\ .1 & .7 & .25 \\ .1 & .3 & .5 \end{pmatrix}.$$

In the long run, 35.7% are at Duke, 35.7% are at UNC and 28.6% are at NC State.

3. $P(Duke|Durham)$

$= \frac{P(Durham|Duke)P(Duke)}{P(Durham|Duke)P(Duke)+P(Durham|UNC)P(UNC)+P(Durham|NCState)P(NCState)}$

$= \frac{.4*.5}{.4*.5+.1*.2+.2*.3} \approx .71$

4. $T = \begin{pmatrix} .75 & 0 & .7 \\ .25 & .8 & 0 \\ 0 & .2 & .3 \end{pmatrix}$ and $E = \begin{pmatrix} .4 & .5 & .1 \end{pmatrix}$.

a. $P(O_1 = Sad) = 0.2955$

b. $P(O_1 = Sad, O_2 = Sad) = 0.10275$

c. $P(O_1 = Sad, O_2 = Sad, O_3 = Happy) = 0.0646725$

d. $\frac{.0261}{P(\{Sad, \ Sad, \ Happy\})} = 0.403572$

e. $\{Bull, Bull, Bull\}$

6. From Example 2.16, $\pi = \{.91, .09\}$ and

$$T = \begin{pmatrix} 0.74914 & 0.52723 \\ 0.25086 & 0.47277 \end{pmatrix}, E = \begin{pmatrix} 0.678328 & 0.3725 \\ 0.321672 & 0.6275 \end{pmatrix}.$$

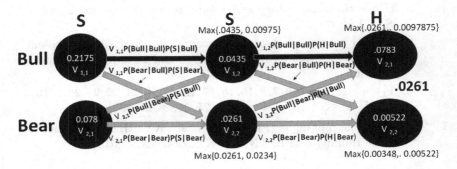

5.

FIGURE 8.1
Recursive decisions in a Viterbi Algorithm Matrix for Exercise 5.

$\alpha(H_1) = \{0.6170, 0.0337\}$	$\beta(H_1) = \{0.0649, 0.0527\}$
$\alpha(H_2) = \{0.1544, 0.1071\}$	$\beta(H_2) = \{0.2303, 0.0555\}$
$\alpha(H_3) = \{0.1168, 0.0333\}$	$\beta(H_3) = \{0.3692, 0.1193\}$
$\alpha(H_4) = \{0.0713, 0.0168\}$	$\beta(H_4) = \{0.3984, 0.4663\}$
$\alpha(H_5) = \{0.02, 0.0162\}$	$\beta(H_5) = \{1, 1\}$

So

and the new model is λ with $\pi^{new} = \{0.957557, 0.0424426\}$

$$T^{new} = \begin{pmatrix} 0.730663 & 1.11403 \\ 0.17552 & 0.563931 \end{pmatrix}, E^{new} = \begin{pmatrix} 0.653434 & 0.367311 \\ 0.346566 & 0.632689 \end{pmatrix}.$$

7. $\alpha(H_1) = \{0.1, 0.05625, 0.1, 0.063\}, \alpha(H_2) = \{0.0271, 0.0093, 0.0189, 0.0297\}$, and $\alpha(H_3) = \{0.0042, 0.0051, 0.0032, 0.0043\}$. $P(O|\lambda) = 0.017$

8. $\gamma(H_1) = \{0.3633, 0.0846, 0.29, 0.2622\}$, $\gamma(H_2) = \{0.3145, 0.121, 0.2067, 0.3578\}$ and $\gamma(H_3) = \{0.25, 0.3036, 0.1905, 0.256\}$. So the most likely hidden sequence is **ATC**.

8.3 Chapter 3

1. a. The transformed vectors will be $\{1,2\}$ and $\{3,4\}$. These vectors are not orthogonal.

   b. $\theta \approx 0.9569 \pm 2\pi n, 2.5277 \pm 2\pi n, 4.0985 \pm 2\pi n, 5.6693 \pm 2\pi n$, for $n \in \mathbf{Z}$.

c. Using $\theta = 0.9569$, $u_1 = \{2.21088, 4.99781\}$ and $u_2 = \{0.334681, -0.148053\}$. The singular values are $\sigma_1 = ||u_1|| = 5.46499$ and $\sigma_2 = ||u_2|| = 0.365966$.

d. $AA^T = \begin{pmatrix} 5 & 11 \\ 11 & 25 \end{pmatrix}$ has eigenvalues $15 \pm \sqrt{221}$. Thus the singular values of A are $\sqrt{15 \pm \sqrt{221}}$.

2. The singular values are 2 and 5.

3. The singular values and eigenvalues are 6.0216, 2.69854, and 0.676943.

4. The singular values of A are the square root of the eigenvalues of $A^T A$. Let λ be an eigenvalue of A, then λ is also an eigenvalue of A^T. If v is the eigenvector corresponding to eigenvalue λ, then $A^T A x = \lambda A^T x = \lambda^2 x$. Thus λ is a singular value of A.

5. a.

$$U = \begin{pmatrix} 0.476971 & -0.34947 \\ 0.187933 & -0.548739 \\ 0.389186 & 0.476451 \\ 0.281154 & -0.48811 \\ 0.533879 & 0.153706 \\ 0.470786 & 0.296436 \end{pmatrix}.$$

A graph would use the rows of U and the (x,y) point for each type of dog.

b. Breed 4 is closest to Carl's dog.

6. a. If the data is normalized, PCA on the covariance matrix produces the model

$$\frac{\widehat{population} - 9199.86}{2619.25} = -0.707107 \frac{hectares - (5.28422 * 10^6)}{718594} + 0.707107$$

or

$$\widehat{population} = -0.00257738(hectares) + 24671.4.$$

7. See GitHub links 19 and 21.

8. Notice that there are really three distinct clusters in Figure 8.2 of stories based on the attributes that are present.

9. a. Figure 8.3

b. William & Mary

c. University of Richmond

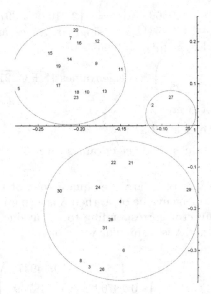

FIGURE 8.2
Visualization for Exercise 8 (Chapter 3).

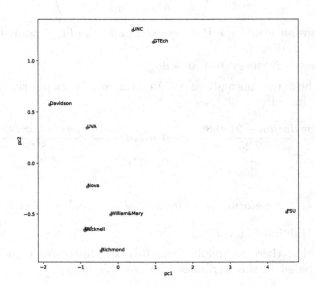

FIGURE 8.3
Visualization for Exercise 9a (Chapter 3).

d.

$$S_w = \begin{pmatrix} 26.9399 & 1.15269 \\ 1.15269 & 4.27101 \end{pmatrix}, S_b = \begin{pmatrix} 0.372925 & -0.461075 \\ -0.461075 & 0.570062 \end{pmatrix}$$

e. $\{-0.0305852, 0.185033\}$

8.4 Chapter 4

1. $\hat{f}(x) = 11255.1 - 957.206x + 29.7221x^2 - 0.396603x^3 + 0.00191023x^4$

2. $P_4(x) = \frac{-168x^4 + 5321x^3 - 54372x^2 + 208279x - 205260}{9240}$

3. $P_5(x) = \frac{1}{120}(-51522x^5 + 897185x^4 - 5520160x^3 + 14609515x^2 - 9915458x + 7828080)$

4. $T_4(x) = 1 - 8x^2 + 8x^4$ and $T_5(x) = 5x - 20x^3 + 16x^5$.

$$\int_{-1}^{1} (1 - x^2)^{-1/2} T_4(x) T_5(x) = 0.$$

5. $U_3(x) = x^3 - 2x$ and $U_4(x) = 1 - 3x^2 + x^4$.
$\int_{-1}^{1}(1 - x^2)^{-1/2}U_3(x)U_4(x)dx = 0.$

6. $\int_{-\infty}^{\infty}(2x)(4x^2 - 2)e^{-x^2/2}dx = 0$

7. $L = \begin{pmatrix} 1 & 0 & 0 \\ 1 & 1 & 0 \\ 1 & -2 & 1 \end{pmatrix}, U = \begin{pmatrix} 1 & 1 & 2 \\ 0 & 1 & 3 \\ 0 & 0 & 7 \end{pmatrix}$

8. a. $\begin{pmatrix} 1 & 1 & 1 & 1 & 1 & 1 \\ 1 & 2 & 4 & 8 & 16 & 32 \\ 1 & 3 & 9 & 27 & 81 & 243 \\ 1 & 4 & 16 & 64 & 256 & 1024 \\ 1 & 5 & 25 & 125 & 625 & 3125 \\ 1 & 6 & 36 & 216 & 1296 & 7776 \end{pmatrix}$

b. $L = \begin{pmatrix} 1 & 0 & 0 & 0 & 0 & 0 \\ 1 & 1 & 0 & 0 & 0 & 0 \\ 1 & 2 & 1 & 0 & 0 & 0 \\ 1 & 3 & 3 & 1 & 0 & 0 \\ 1 & 4 & 6 & 4 & 1 & 0 \\ 1 & 5 & 10 & 10 & 5 & 1 \end{pmatrix}, U = \begin{pmatrix} 1 & 1 & 1 & 1 & 1 & 1 \\ 0 & 1 & 3 & 7 & 15 & 31 \\ 0 & 0 & 2 & 12 & 50 & 180 \\ 0 & 0 & 0 & 6 & 60 & 390 \\ 0 & 0 & 0 & 0 & 24 & 360 \\ 0 & 0 & 0 & 0 & 0 & 120 \end{pmatrix}.$

  c. $\{65397, 59438, -5990, \frac{2569}{3}, \frac{24871}{24}, -\frac{8587}{20}\}$

  d. $\frac{1}{120}(-51522x^5 + 897185x^4 - 5520160x^3 + 14609515x^2 - 9915458x + 7828080)$

10.   a. If $t = 0$ in year 2016, then

$$f(\hat{x}) = -\frac{24201x^3}{4} + \frac{300931x^2}{7} - \frac{1910463x}{28} + \frac{8909809}{70}.$$

  b. Approximate derivatives

$$\{\{2016, -18436\}, \{2017, -18436\}, \{2018, -7322\}, \{2019, 57475\}, \{2020, -4014\}\}.$$

8.5   Chapter 5

1.

$$\begin{pmatrix} 2x_1 & 1 & 0 \\ 0 & 8x_2 & 3x_3^2 \end{pmatrix}$$

2. $\nabla f = \{y\cos(xy + y^2 z), (x + 2yz)\cos(xy + y^2 z), y^2 \cos(xy + y^2 z)\}$

$$\frac{\partial^2 f}{\partial x^2} = -y^2 \sin(xy + y^2 z)$$

$$\frac{\partial^2 f}{\partial y^2} = 2z\cos(xy + y^2 z) - (x + 2yz)^2 \sin(xy + y^2 z)$$

$$\frac{\partial^2 f}{\partial z^2} = -y^4 \sin(xy + y^2 z)$$

$$\frac{\partial^2 f}{\partial y \partial x} = \frac{\partial^2 f}{\partial x \partial y} = \cos(xy + y^2 z) - y(x + 2yz)\sin(xy + y^2 z)$$

$$\frac{\partial^2 f}{\partial y \partial z} = \frac{\partial^2 f}{\partial z \partial y} = 2y\cos(xy + y^2 z) - y^2(x + 2yz)\sin(xy + y^2 z)$$

$$\frac{\partial^2 f}{\partial z \partial x} = \frac{\partial^2 f}{\partial x \partial z} = -y^3 \sin(xy + y^2 z)$$

3. $\nabla f = \{8x_1 + 6x_2^3, 18x_1 x_2^2\}$.

4. $d(ln(|X|)) = tr(X^{-1} dX)$

5. $dX^T = (dX)^T$

6.   a.

$$LL = \sum_{i=1}^{n} log(y_i - (a_0 + a_1 x_i)) - log(24) - \frac{(y_i - (a_0 + a_1 x_i))^2}{2}$$

b.
$$\frac{\partial LL}{\partial a_0} = \sum_{i=1}^{n} \frac{-1}{y_i - (a_0 + a_1 x_i)} - (y_i - (a_0 + a_1 x_i))$$

c.
$$\frac{\partial LL}{\partial a_1} = \sum_{i=1}^{n} \frac{-x_i}{y_i - (a_0 + a_1 x_i)} - x_i(y_i - (a_0 + a_1 x_i))$$

7. Since XA not necessarily invertible, we create $(XA)^T(XA)$,

$$\frac{\partial f}{\partial a_0} = (XA)^T((XA)^T(XA))^{-1}Y.$$

8.
$$\frac{\partial |a_0 + a_1 x_i - y_1|}{\partial a_0} = \begin{cases} 1 & \text{if } a_0 + a_1 x_i \geq 0, \\ -1 & \text{if } a_0 + a_1 x_i < 0. \end{cases}$$

$$\frac{\partial |a_0 + a_1 x_i - y_1|}{\partial a_1} = \begin{cases} x_i & \text{if } a_0 + a_1 x_i \geq 0, \\ -x_i & \text{if } a_0 + a_1 x_i < 0. \end{cases}$$

9.   a. See Table 8.1.

TABLE 8.1

Gradient descent table for Exercise 9a (Chapter 5).

y_i	$\hat{y}_i = a_0 + a_1 x_i$	$\frac{\partial L}{\partial a_0}$	$\frac{\partial L}{\partial a_1}$
3	5.5	1	-1
-4	5	1	0
5	4.5	-1	-1
2	4	1	-6
Total		2	-8

b. $a_0 = 5 - .03(2) = 4.94, a_1 = -.5 - .03(8) = -.74$.

10. See Table 8.2. With a learning rate of .03, $a_0 = 0 - .03(-2) = .06, a_1 = 0 - .03(-2) = .06$.

11.   a. $SSE = 500783828$

b. Using the quadratic interpolating function, we get gradient Table 8.3.

c. With a learning rate of $r = .0003$, $f(\hat{x}) = 39991.1 + 29961.x + 3808.73x^2$ and $SSE = 4.63776 \times 10^8$

12. $\hat{y} = \frac{9}{5} + \frac{3}{10}x$

TABLE 8.2
Gradient descent table for Exercise 9b (Chapter 5).

y_i	$\hat{y}_i = a_0 + a_1 x_i$	$\frac{\partial L}{\partial a_0}$	$\frac{\partial L}{\partial a_1}$
3	0	-1	1
-4	0	1	0
5	0	-1	-1
2	0	-1	-2
Total		-2	-2

TABLE 8.3
Gradient descent table for Exercise 11b (Chapter 5).

							Total
x	1	2	3	4	5	6	
y=Total Infections	65397	124835	172293	212909	276692	367000	
$\hat{y}$	74000	116000	166000	224000	290000	364000	
$\frac{\partial SSE}{a_0}$	17206	-17670	2586	22182	26616	-6000	29748
$\frac{\partial SSE}{a_1}$	17206	-35340	-37758	88728	133080	-36000	129916
$\frac{\partial SSE}{a_2}$	17206	-70680	-113274	354912	665400	-216000	637564

13.   a. $\dfrac{1}{1 + e^{-XW}} = \dfrac{1}{1 + e^{.1+.9x}}$

      b. $SSE = 0.462321$

      c. Gradients are $\{0.424522, 0.155872\}$, $W^{(1)} = \{0.0872644, 0.895324\}$ and $SSE = 0.456202$.

      d. $ReLU(XW) = XW = \{0.1, 1., 1.9\}$ since all values XW are positive.

      e. $SSE = 1.5$

      f. Gradients are $\{3.6, 5.4\}$ and $W = \{0.064, 0.846\}$.

8.6   Chapter 6

1.   a. *entropy* $\approx .52$, $IG = 1 - \frac{7}{10} \cdot .52 \approx .36$.

      b. One example is $y > -.2$.

2. One example is that $\hat{y} = 0.038988$, when $x < -7.4495$ and $\hat{y} = 0.164017$, when $x \geq -7.4495$

3. b. See Figure 8.4.

    c. Sample root criteria $x > 1.5$ and child $y > 1.5$

    d. With root criteria $x > 1.5$ and child $y > 1.5$, Sri Lanka is predicted to have a low substance abuse measure in 2019. If one takes the average of the training data in this region the value is 0.383333.

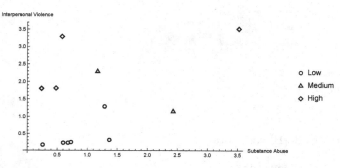

FIGURE 8.4
Visualization for Exercise 3b (Chapter 6).

8.7   Chapter 7

1. $\kappa(A) = 20.0501$.

5. Matrices a and c are Hermitian.

6. If $r_1 = \{\frac{1+i}{2}, -\frac{1+i}{2}\}$ and $r_2 = \{\frac{1}{\sqrt{2}}, \frac{1}{\sqrt{2}}\}$. Then $||r_1|| = ||r_2|| = 1$ and $r_1.r_2 = 0$. Similarly $c_1 = \{\frac{1+i}{2}, \frac{1}{\sqrt{2}}\}$ and $c_2 = \{-\frac{1+i}{2}, \frac{1}{\sqrt{2}}\}$ then $||c_1|| = ||c_2|| = 1$ but $c_1.c_2 \neq 0$.

7. (Sample Python Code, below, and Sample distribution in Figure 8.5)
    ```
    import matplotlib.pyplot as plt
    import numpy as np
    import random
    values= random.sample(range(100, 1000), 500)
    data = np.zeros((500,365))
    tdata =np.transpose(data)
    for time in range(0,365):
    for i in range(0,100):
    trade=random.sample(range(0,500),2)
    if values[trade[0]]¿=values[trade[1]]:
    values[trade[0]]=.75*total
    ```

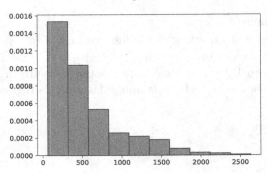

FIGURE 8.5
Distribution of values.

```
values[trade[1]]=.25*total
else:
values[trade[0]]=.25*total
values[trade[1]]=.75*total
tdata[time]=values
data=np.transpose(tdata)
evalues=np.linalg.eig(np.dot(data,tdata))
plt.hist(values, bins=10, density=True, color = "gray"',edgecolor =
'black')
```

Github Links

1. https://github.com/ccoles1/LinearAlgebra_datasets/blob/master/cinderellastories.csv

2. https://github.com/ccoles1/LinearAlgebra_datasets/blob/master/pythonchapter1.py

3. https://github.com/ccoles1/LinearAlgebra_datasets/blob/master/project1sports.csv

4. https://github.com/ccoles1/LinearAlgebra_datasets/blob/master/ncaamens2020a.csv

5. https://github.com/ccoles1/LinearAlgebra_datasets/blob/master/ncaamens2020b.csv

6. https://github.com/ccoles1/LinearAlgebra_datasets/blob/master/pythonchapter1b.py

7. https://github.com/ccoles1/LinearAlgebra_datasets/blob/master/rchapter1.R

8. https://github.com/ccoles1/LinearAlgebra_datasets/blob/master/rchapter1b.R

9. https://github.com/ccoles1/LinearAlgebra_datasets/blob/master/weightedcensussmall.csv

10. https://github.com/ccoles1/LinearAlgebra_datasets/blob/master/censusall.csv

11. https://github.com/ccoles1/LinearAlgebra_datasets/blob/master/pythonchapter2b.py

12. https://github.com/ccoles1/LinearAlgebra_datasets/blob/master/rchapter2b.R

13. https://github.com/ccoles1/LinearAlgebra_datasets/blob/master/bigram.csv

14. https://github.com/ccoles1/LinearAlgebra_datasets/blob/master/pythonchapter2c.py

15. https://github.com/ccoles1/LinearAlgebra_datasets/blob/master/rchapter2c.R

16. https://github.com/ccoles1/LinearAlgebra_datasets/blob/master/s_n_p_data1920.csv

17. https://raw.githubusercontent.com/rfordatascience/tidytuesday/master/data/2020/2020-01-21/spotify_songs.csv

18. https://github.com/ccoles1/LinearAlgebra_datasets/blob/master/tweets.csv

19. https://github.com/ccoles1/LinearAlgebra_datasets/blob/master/pythonchapter3a.py

20. https://github.com/ccoles1/LinearAlgebra_datasets/blob/master/pythonchapter3b.py

21. https://github.com/ccoles1/LinearAlgebra_datasets/blob/master/rchapter3a.R

22. https://github.com/ccoles1/LinearAlgebra_datasets/blob/master/rchapter3b.R

23. https://github.com/ccoles1/LinearAlgebra_datasets/blob/master/starhusbandbook.csv

24. https://github.com/ccoles1/LinearAlgebra_datasets/blob/master/faces.csv

25. https://github.com/ccoles1/LinearAlgebra_datasets/blob/master/fullfaces.csv

26. https://github.com/ccoles1/LinearAlgebra_datasets/blob/master/pythonchapter4a.py

27. https://github.com/ccoles1/LinearAlgebra_datasets/blob/master/pythonchapter4b.py

28. https://github.com/ccoles1/LinearAlgebra_datasets/blob/master/pythonchapter4c.py

29. https://github.com/ccoles1/LinearAlgebra_datasets/blob/master/rchapter4a.R

30. https://github.com/ccoles1/LinearAlgebra_datasets/blob/master/rchapter4b.R

31. https://github.com/ccoles1/LinearAlgebra_datasets/blob/master/rchapter4c.R

32. https://github.com/ccoles1/LinearAlgebra_datasets/blob/master/pythonchapter5a.py

33. https://github.com/ccoles1/LinearAlgebra_datasets/blob/master/pythonchapter5b.py

34. https://github.com/ccoles1/LinearAlgebra_datasets/blob/master/rchapter5a.R

35. https://github.com/ccoles1/LinearAlgebra_datasets/blob/master/rchapter5b.R

36. https://github.com/ccoles1/LinearAlgebra_datasets/blob/master/digits.csv

37. https://github.com/ccoles1/LinearAlgebra_datasets/blob/master/pythonchapter6a.py

38. https://github.com/ccoles1/LinearAlgebra_datasets/blob/master/pythonchapter6b.py

39. https://github.com/ccoles1/LinearAlgebra_datasets/blob/master/rchapter6a.R

40. https://github.com/ccoles1/LinearAlgebra_datasets/blob/master/rchapter6b.R

41. https://github.com/ccoles1/LinearAlgebra_datasets/blob/master/wordle.py

42. https://github.com/ccoles1/LinearAlgebra_datasets/blob/master/words.csv

43. https://github.com/ccoles1/LinearAlgebra_datasets/blob/master/birdcallsupdated.zip

44. https://github.com/ccoles1/LinearAlgebra_datasets/blob/master/pythonchapter7a.py

45. https://github.com/ccoles1/LinearAlgebra_datasets/blob/master/pythonchapter7b.py

46. https://github.com/ccoles1/LinearAlgebra_datasets/blob/master/rchapter7a.R

47. https://github.com/ccoles1/LinearAlgebra_datasets/blob/master/rchapter7b.R

48. https://github.com/ccoles1/LinearAlgebra_datasets/blob/master/stockreturn.zip

49. https://github.com/ccoles1/LinearAlgebra_datasets/blob/master/GSE13598.txt

50. https://github.com/ccoles1/LinearAlgebra_datasets/blob/master/diseasedatabase.csv

Bibliography

[1] ACLU. *Extreme Racial Disparities Persist in Marijuana Arrests.* Retrieved from https://graphics.aclu.org/marijuana-arrest-report/.

[2] Barnett, N. (2018). *NC Case Could Slay Dragon of Partisan Gerrymandering.* News and Observer.

[3] Benson-Putnins, D., Bonfardin, M., Magnoni, M. E., Martin, D. (2011). *Spectral Clustering and Visualization: a Novel Clustering of Fisher's Iris Data Set.* SIAM Undergraduate Research Online, 4, 1–15.

[4] Boyle, M. (2015). *Notes on the Perron-Frobenius Theory of Nonnegative Matrices.* Dept. of Mathematics, University of Maryland, College Park, MD. USA, Retrieved from http://www. math. umd. edu/ mboyle/courses/475sp05/spec. pdf.

[5] The Cancer Network Galaxy. Database of Cancer Gene Networks from Public Gene Expression Data. Retrieved from https://tcng.hgc.jp/

[6] Chartier, T., Kreutzer, E., Langville, A., Pedings, K. (2010). *Bracketology: How Can Math Help?.* Mathematics and Sports, 43(67), 55–70.

[7] Cushing, J. M., Yicang, Z. (1994). *The Net Reproductive Value and Stability in Matrix Population Models.* Natural Resource Modeling, 8(4), 297–333.

[8] De Abreu, N. M. M. (2007). *Old and New Results on Algebraic Connectivity of Graphs.* Linear algebra and its applications, 423(1), 53–73.

[9] Dundes, A. (editor). (1965). *Introduction to Stith Thompson's "The star husband tale".* pp. 414–415. Prentice-Hall. Englewood Cliffs, NJ.

[10] Evans, M. L., Lindauer, M., Farrell, M. E. (2020). *A Pandemic within a Pandemic—Intimate Partner Violence during Covid-19.* New England journal of medicine, 383(24), 2302–2304.

[11] Fernendez, J. (2018). *Judges Rule N.C.'s 2016 Congressional Districts Partisan Gerrymander.* Winston-Salem Journal.

[12] Fiedler, M. (1975). *A Property of Eigenvectors of Nonnegative Symmetric Matrices and its Application to Graph Theory.* Czechoslovak Mathematical Journal, 25(4), 619–633.

[13] Yahoo! Finance. Retrieved from finance.yahoo.com

[14] Free Images. Retrieved from https://www.freeimages.com/search/face

[15] Gillman, R. (2002). *Geometry and Gerrymandering.* Math Horizons, 10(1), 10-12.

[16] Gupta, A. and Dhingra, B. (2012). *Stock Market Prediction using Hidden Markov Models.* In 2012 Students Conference on Engineering and Systems (pp. 1–4). IEEE.

[17] Heiner, H. A. (Ed.). (2012). *Cinderella Tales from Around the World: Fairy Tales, Myths, Legends and Other Tales with Cinderellas.* SurLaLune Press.

[18] American Council on Education. Data Tables. Retrieved from https://www.equityinhighered.org/data-tables/.

[19] Keener, J. P. (1993). *The Perron–Frobenius Theorem and the Ranking of Football Teams.* SIAM review, 35(1), 80–93.

[20] Kershenbaum, A. (2014). *Entropy Rate as a Measure of Animal Vocal Complexity.* Bioacoustics, 23(3), 195–208.

[21] Liu, E. (2019) *Yes, Trump's Tweets Move the Stock Market. But Not for Long.* Barron's. Retrieved from https://www.barrons.com/articles/donald-trump-twitter-stock-market-51567803655

[22] Minters, M. (2018) *Diverse: Issues In Higher Education Names Top 100 Degree Producers.* Diverse Issues in Higher Education. Retrieved from https://diverseeducation.com/article/120916/

[23] National Domestic Hotline. *Impact and State Reports.* Retrieved from https://www.thehotline.org/stakeholders/impact-and-state-reports/

[24] National Coalition Against Domestic Violence (NCADV). (2021). *COVID-19 and Domestic Violence* Retrieved from https://ncadv.org/covid-19-and-domestic-violence

[25] Nguyen, N. (2018). *Hidden Markov Model for Stock Trading.* International Journal of Financial Studies, 6(2), 36.

[26] North Carolina General Assembly. *2011 Redistricting Reference Data.* Retrieved from https://www.ncleg.gov/Redistricting/BaseData2011

[27] Ontario (2020) Ontario Government. (2020). Confirmed Positive Cases of COVID-19 in Ontario. [Data file]. Retrieved from https://data.ontario.ca/dataset/f4112442-bdc8-45d2-be3c-12efae72fb27/resource/455fd63b-603d-4608-8216-7d8647f43350/download/conposcovidloc.csv.

[28] Pierce, J. R. (2012). *An Introduction to Information Theory: Symbols, Signals and Noise.* Courier Corporation.

[29] Ralston, A. (2008). *Gene Interaction and Disease.* Nature Education 1(1), 16.

[30] Trump Twitter Archive V2. (n.d.). Retrieved from https://www. thetrumparchive.com/

[31] Slininger, B. (2013). *Fiedlers Theory of Spectral Graph Partitioning.* Dosegljivo: https://citeseerx.ist.psu.edu/document?repid=rep1&type= pdf&doi=beeee28a9c52287ac9a20868d067d1ed20788036.

[32] Statista. Poverty Rates in the United States 1990-2020. Retrieved from https://www.statista.com/statistics/200463/us-poverty-rate-since-1990/

[33] US Department of Education. (2016). Advancing diversity and inclusion in higher education. Retrieved from https://www2.ed.gov/rschstat/ research/pubs/advancing-diversity-inclusion.pdf

[34] Urschel, J. C., Zikatanov, L. T. (2014). *Spectral Bisection of Graphs and Connectedness.*Linear Algebra and its Applications, 449, 1–16.

[35] Violence Policy Center. (2018). State Firearm Death Rates, Ranked by Rate, 2018. It sources the National Center for Injury Prevention and Control of the Centers for Disease Control and Prevention. Retrieved from http://vpc.org/state-firearm-death-rates-ranked-by-rate-2018/

[36] Von Luxburg, U. (2007). *A Tutorial on Spectral Clustering.* Statistics and Computing, 17(4), 395–416.

[37] Wey, T. W., Blumstein, D. T. (2010). *Social Cohesion in Yellow-Bellied Marmots is Established through Age and Kin Structuring.* Animal Behaviour, 79(6), 1343–1352.

[38] Wilson, A. G., Ghahramani, Z. (2010). *Generalised Wishart processes.* arXiv preprint arXiv:1101.0240.

[39] Wines, W. (2019). *State Court Bars Using North Carolina House Map in 2020 Elections.* New York Times. Retrieved from https://www.nytimes.com/2019/10/28/us/north-carolina-gerrymander-maps.html.

[40] Ritchie, H. and Roser M. (n.d.). "Our World in Data". Retrieved from https://ourworldindata.org/energy#energy-country-profiles

[41] The World Bank, GDP (U.S.D) Data. (n.d.) Retrieved from https://data.worldbank.org

[42] World Population Review. (n.d.) Retrieved from https:// worldpopulationreview.com/states/gun-ownership-by-state/

[43] xeno-canto, Sharing Bird Songs From around the World. (n.d.) Retrieved from https://xeno-canto.org/

[44] LaCunn, Y., Cortes, C., and Burges, C. (n.d.) "The MNIST Database for handwritten digits". Retrieved from http://yann.lecun.com/exdb/mnist/

[45] Young, F.W. (1978). *Folktales and Social Structure: a Comparison of Three Analyses of the Star-Husband Tale.* J. Amer. Folklore 91(360), 691–699.

Index

Printed in the United States
by Baker & Taylor Publisher Services